REAL LIFE
CONSTRUCTION
MANAGEMENT GUIDE
FROM A-Z

JAMIL SOUCAR

ISBN: 979-8-88640-595-8 (sc)
ISBN: 979-8-88640-269-8 (hc)
ISBN: 979-8-88640-597-2 (e)

One Galleria Blvd., Suite 1900, Metairie, LA 70001
1-888-421-2397

Dedication

I would like to dedicate this new version of the
book to my wife and twins, my son and daughter, who
contributed to the new design of the book
and for their support.
A special dedication to my wife, who spent many
days on the edit review, working from her heart
with her genuine care to help me republish a better
version of this book.

CONTENTS

About This Book !

The reason I decided to write this book is to provide people in the construction industry a different perspective on the various aspects relevant to managing construction projects.

You probably have read several books, whether in college or for your own additional education, to learn and improve your skills in managing construction projects. There are so many books in the market that teach the principles of scheduling, cost control, construction documentation, change order, etc. I have confidence that all these books have done a great job outlining the technical concepts involved. However, I haven't come across a book that taught me about the real-life pitfalls that usually happen when implementing these concepts.

This book is a bit different. My intent is to go beyond the typical book format that sounds like a college textbook teaching core concepts. In this book, I would like to talk to you. I want to tell you about the real-life and human factors that impact how to best run projects. In addition to explaining the theory behind each topic, this book will offer you a guideline for best practices.

Human nature, personalities, and egos are big factors in managing construction projects. The physical construction is the easiest part of the fabric of a project. The challenge is how to navigate through the multiple personalities, politics, and agendas, not to mention avoiding litigation in an industry that is very litigious.

A simple Request for Information (RFI) written the wrong way can intimidate the architect or the owner and change the tone of the project. I have always told my students that construction management is 20 percent construction and 80 percent people skills.

In this book, I share with you my experience and lessons learned the hard way. Most of the chapters of this book are divided into the following sections for each topic:

• Theories that teach the concepts of various construction management topics

• Real- Life Gauge perspective that explains how things work in real life

• Best Practices Tips that give you a summarized set of tips for best practices

Hope you enjoy the book.

Chapter 1
The Undisclosed Factors

The main message of this book is that the human factors and personalities are big factors in managing construction projects. I published this article a few years back, and I thought it would make a good start for your reading to set the stage for that main theme of this book.

We often hear about various problems that take place on construction projects. Most common problems include the following: A bad set of plans, a contractor not performing well, delays, unforeseen conditions, change orders, etc. These are legitimate and real issues that we have to deal with.

However, I would like to talk about the undisclosed causes that often contribute to the problems and disputes. These are factors that the project team often does not disclose or admit exist. This is what I am referring to as the "undisclosed factors".

Construction project teams include a wide variety of people, including but not limited to owners, architects, construction managers, inspectors, community members, city officials, general contractors, subcontractors, and sometimes more parties. If we assume that the measure of success for all these people is the success of the project, shouldn't you logically assume that this common goal is enough to make everyone work together and help each other out? Obviously, that is not the case a lot of times. Often, we get disputes and claims. Sometimes there are legitimate reasons to these conflicts, and sometimes it is because of the "undisclosed factors." Please read on.

Many years ago, I was working as a project manager for a general contractor. We were at one of our weekly site meetings discussing a construction problem that needed a resolution. All the above-mentioned parties were present at that time. Each one contributed his or her feedback to resolve the problem, and gradually, the atmosphere became tense and adversarial. While I was silently listening to the various proposed solutions, I was privately thinking, *"The solution to this problem is so simple, and I can get it done if only these people will leave us alone and go back to their offices"*. I went on to think, *"why are they complicating this "*

Well, this was when I realized that each one had a different agenda, a different set of fears and liabilities, and each one was trying to push a solution that fit his/her agenda, and not necessarily the best solution to the problem.

On another project, I saw an inspector giving one of the subcontractors an unusually tough time passing the requested inspection. They almost got into a fistfight. I later took the inspector aside and asked him why. His answer was not technical or referring to a specification section, despite the fact that the loud argument was all about specs and plans. His answer was simply, "This sub doesn't talk to me with respect".

Sometimes, conflicts on site arise because of personality conflicts between the parties. Again, no one admits that, but instead, they look through the contract documents to find clauses that may give their position a contractual justification if interpreted a certain way.

Most of those people do not admit what is described above; instead, the conflict grows larger. That is why I call them the "undisclosed factors."

Managing a construction project requires a skill in managing personalities and characters of the involved parties in addition to the required technical knowledge.

Relationship counselors often teach that when you enter into a relationship with a person, you are taking in the person plus what they call "baggage," which is all his/her background, flaws, strengths, and pasts. The same applies to the project's team that just met and has to interact with each other on a daily basis for the duration of the project. Perhaps project specifications should allocate a budget for a relationship psychologist. A wild idea, but it may help save money by having fewer claims.

I have managed a lot of projects throughout my career from both sides: the contractor's and the owner's. What I concluded is that to have a smooth-running project, there is no substitute for building relationships based on fairness and flexibility, focusing on solving problems rather than placing blame, being practical, and promoting what works. All parties have money at risk. Make sure that all parties are profitable.

I certainly do not mean to paint a bad picture about the professionals managing projects. In fact, throughout my career, I have had the good fortune to interact with better helpful professionals than the ones described above. I can write a whole new article citing examples of great inspectors, architects, and construction managers who helped me resolve problems and brought projects to a successful completion. I'm simply trying to raise awareness about the above factors that do exist and should be kept in perspective.

Different people may look at a project like how the people below are looking at the elephant. Your job as the construction manager is to manage all the players to make them see this as an elephant.

Chapter 2
History of Construction Management

Construction management had its beginnings in the early to mid-1960s. At that time, the need for capital facilities and improvements in the marketplace was in great demand. Additionally, major technology and methodology developments were occurring and played a significant role in the evolution of construction management. For example, the electronic computer and sophisticated scheduling methods like the Construction Project Manager (CPM).

One of the earliest reported projects to use construction management was New York City's Madison Square Garden in 1963. Other projects were:

- The 100-story John Hancock Center in Chicago
- The twin 110-story towers of New York's World Trade Center
- John Manville World Headquarter building in Denver
- First National City Bank building in New York

Efforts were made to establish standards for the practice of construction management. It all began with a 1970 study by the General Services Administration (GSA) of its construction methods on public building services construction contracting. GSA wanted to capitalize on the experiences of the private sector and to change the way it had been doing its design and construction business. Their report recommended that GSA abandon its outmoded procedures and use phased construction in conjunction with construction management in a new dynamic approach to its nationwide building program.

This system was tried in several projects. GSA abandoned the use of construction management in the latter part of 1979 based on their belief that the benefits expected by GSA were not being achieved.

With the advent of construction management and the rapid trend among owners toward using it, in 1972, the Associated General Contractors of America (AGC) adopted guidelines for its suggested approach to construction management. This was followed by the development of a family of standard contract forms for use by its membership. In establishing these guidelines, AGC maintained that it did not endorse construction management as a substitute for any other successful contracting format. The American Institute of Architects (AIA) likewise embraced construction management as a viable alternative, and they too published a family of construction management documents in 1976.

In August 1975, the National Construction Management Committees of AGC, AIA, and the American Consulting Engineers Council (ACEC) met for a joint national construction management council. These associations recognized the importance of the construction management process and agreed to work together on a national comprehensive construction management program.

While the intent and the objectives stated above were sound, this joint effort never became a reality because none of these organizations were primarily devoted to the practice of construction management, and each had its own vested interest in the marketplace.

Founding of the CMAA

In October 1981, representatives of thirty-seven firms practicing construction management met to explore mutual interest in forming a national construction management association. As the result of a strong common interest, the Construction Management Association of America (CMAA) was formed and incorporated in 1982. CMAA provides a unified voice for the construction management industry in those areas where collective representation and action as an association are the best method to achieve common goals.

Here is a summary of the CM history:

- 1960s – CM had its first beginnings.
- 1963 – New York's Madison Square Garden was completed using the CM method.
- 1970 – A study by GSA to introduce CM to public works.
- 1972 – AGC adopted the CM guidelines and developed other forms.
- 1975 – A national CM committee tried to form a national counsel for CM work. This effort failed.
- 1976 – AIA adopted CM guidelines and developed their own forms.
- 1979 – GSA abandoned the use of CM.

- 1981 – Thirty-seven firms practicing CM met to form a national association.
- 1982 – CMAA was formed.

There is a lot to be said about the development of construction management. The main theme is having a construction manager who is a focal point of communication, coordinating and acting as a liaison among the several parties, and mainly facilitating everybody's problems so the project runs smoothly. *Do not forget about the human factor and the undisclosed factors of Chapter 1.*

Chapter 3
The Anatomy of a Construction Project/Key Players

As mentioned in earlier chapters, the main source of difficulty in managing construction projects is the human factor, the various involved parties that come into the project with different perspectives, agendas, and liabilities.

In this chapter, we will introduce the various key players and their typical responsibilities.

Owner: Owns the property, the project, or could be a public agency in public works projects. The owner hires the rest of the players, funds the project, and is usually the final decision maker.

Architect: The main designer of the project hired usually by the owner. The architect is responsible for designing the project, generating the construction documents, and providing technical interpretation and assistance during the construction phase.

Engineers: Engineers are usually hired by the architect. Each specialize in the technical portion of the project. That includes the following:

- Structural Engineer: Designs the structural elements of the building
- Civil Engineer: Designs the exterior site work, including grading
- Electrical Engineer: Designs the electrical portion of the project
- Mechanical Engineer: Designs the air-conditioning and plumbing
- Interior Designer: Designs the interior volume of the building
- Landscape Architect: Designs the landscape and irrigation
- Soil Engineer: Tests the soil and develops the recommendations for proper earth preparation for civil and foundation work.

Inspector(s): Conducts site inspections to verify compliance of work with the contract documents. In some cases, the inspectors work with the city, and in some cases (for example, public schools), the inspector works for the public agency directly. You may have a general inspector who gets trade inspectors to inspect specialty trades like electrical and plumbing.

General Contractor: The prime contractor who is contracted to perform the work. Some execute the work fully or partly with their own workers. Others hire specialty subcontractors.

Subcontractors: Specialty firms like electrical, plumbing, roofing, etc. These firms typically get contracted under the general contractor.

Government Agencies: The party that approves the plans and approves them for construction.

End User: The party that will occupy the facility being built. They should be consulted when developing the scope of work.

Community Groups: Members of the community who have a stake in the project.

Depending on the type of delivery method, the contracts with the owner are different. Below are some examples. We will go over these delivery methods in a later chapter, and we will go into detail about the role of the construction manager in the following chapter.

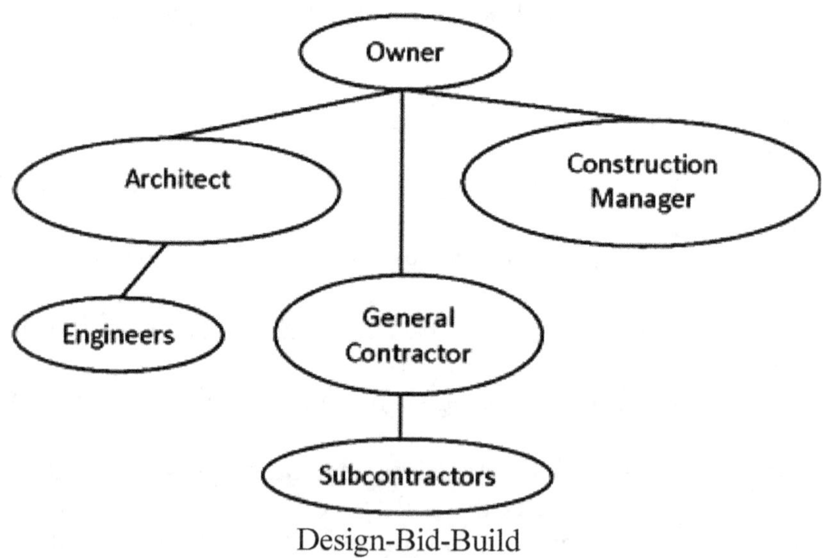

Design-Bid-Build

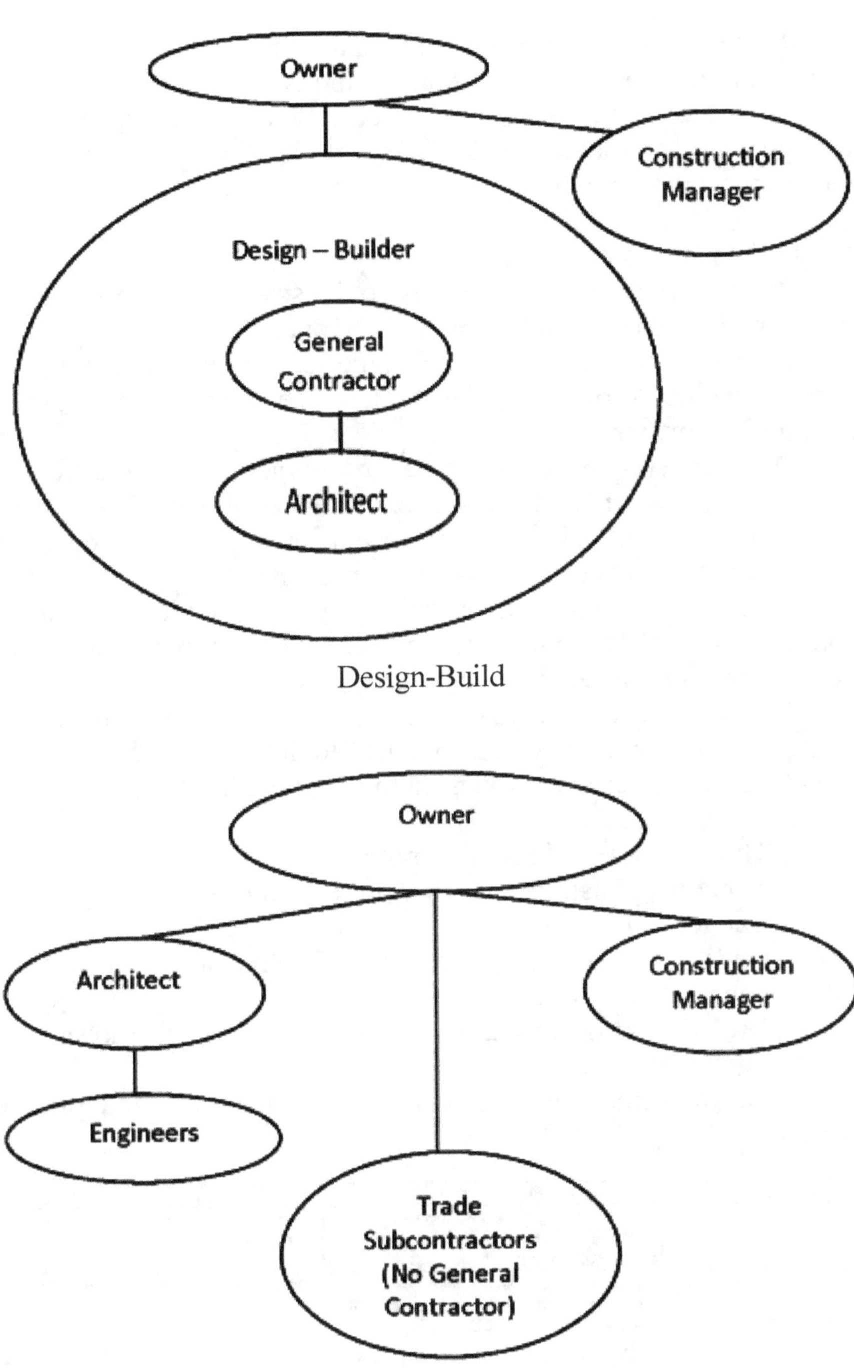

Design-Build

CM/Multiprime

There are a lot more varieties and combinations of contract configurations. The above examples demonstrate the point.

Regardless of who you are, whether you are the construction manager, the contractor, the architect, or others, you should build a good rapport with the owner.

Public vs. Private Owners

Public owners include cities, school districts, counties, or other government agencies funded by public money.

Private owners, as the name suggests, are owners using their money. That could be a homeowner, large developer, or a commercial building owner.

Dealing with these two types of owners is very different. Public works owners and projects are subject to the public contract code and other political considerations that add more complications for and requirements of the construction manager.

With private owners, there is usually more flexibility in decision-making and other administrative tasks as the private owner has more command on decisions regarding the project.

The construction manager has to spend time understanding the owner (the CM's client) to develop the best way of providing his/her service. You need to ask about the following:

- The owner's goals for the project
- Timeline, phasing, and other project milestones
- Budget
- Understand how they process tasks like invoices, change orders, document control, etc.
- Owner's current procedures that the CM is expected to adhere to

Here are some key differences between public and private owners:

Task	Public Owner	Private Owner
Funds	Controlled by the pulic agencies regulations and procedures or other political elements	More control over the funds as it is the Owner's money
Hiring Consultants	Has to issue a public request for proposal, set qualifying criteria where the selection is controlled by public regulations	Can select any consultant they wish
Hiring a Contractor	Has to follow the bid/award rules set by the public contract code	Can hire any firm they wish
Claims	More regulated and less personal	Takes a more personal side

Architect

To have a successful interaction with the architect, it is important to understand the following aspects of his/her work.

When architects give a proposal to the owner to design the project, they base it mostly on the expected number of hours they will spend on the project. If you are an owner or a construction manager, try not to waste their time with unnecessary meetings or continuously change plans during the design phase. If their budget is blown halfway through, then you will be dealing with a tense architect, and the quality of the service and the plans may be impacted.

During construction, the architect's role should be limited to technical support. In case of change order disputes, a lot of architects tend to protect their design and not admit their errors and omissions, which result in claims. That is where the principal of construction management is important; that is being a third party and keeping all players in check.

Inspector

If you are a contractor, then establishing a good working relationship with the inspector is extremely important.

The attitude of the inspector can make or break the project. Referring to Chapter 1, personality clashes may be a huge undisclosed factor hiding behind contract clauses. I have encountered so many different inspectors with different characters. Some of them are a blessing to the project and really help solve problems and keep the work forward, and some of them thrive on rejecting work and placing obstacles in the way.

If you happen to be contractor, then take note of the following tips:

- Respect the inspector's experience and time.
- Do not call them for inspection when the work is not ready. The inspector is not your superintendent.
- Build a trust relationship. I've had inspectors sign off based on my word over the phone.

Some inspectors try to take on more than their role by interfering with the design or means and methods of the contractor. As a construction manager, you need to set the limits of each team members.

General Contractor/Subcontractors

These firms assume the biggest risk of the job. In public works, for example, they have to bid the project so low to be the successful contractor, and sometimes items are missed during the messy bidding process.

If you are an owner or a construction manager, it serves your project well if you assist the builders in getting the job done faster. Be fair and reasonable in responding to inquiries, and be firm in facing some contractors who try to get unjustified money or cut corners. As the construction manager, you need to strike a balance among all team members by defining early on each player's role and the limit of that role.

Unfortunately, in the industry, a lot of people stereotype each other. You will find some construction managers (owner's representatives) who think their role is to deny whatever the contractor claims because contractors are dishonest and every request they make is a game. On the other hand, there are contractors who think that owners will do all they can to leverage their powerful position to deny charges to the contractors. I also have dozens of examples in which I had great interactions between both parties and had great projects.

If you happen to be contractor, remember the following:

- Be a team member and help the owner meet his or her goals.
- Be fair in pricing your change orders.
- Build a trust relationship with the owner, architect, inspector, and others.
- Keep the jobs going no matter what the dispute is. You will be in a better position if you have done what you should have when the other parties are causing the problem.

If you happen to be owner or an owner's representative, take note:

- Your measure of success is in completing the work on time and budget. Since the contractor is the one building the project, that makes it in your best interest to facilitate the contractor's work.
- Stop denying legitimate change order requests because you do not want to pay. Keep it fair and honest. At the same time, be firm facing any contractors' claims that lack merit.
- Keep cash flow to the contractors timely. Nothing hinders the job like lack of cash flow.
- Set a tone of fairness on the project. Other players will follow.

Real-Life Gauge

Obviously, the owner of the project is the main key member of the team. The owner is the party that is hiring the rest of the professionals and paying the bills.

Owners have to understand their important role in setting the tone of the projects. My experience has shown me that the mindset and attitude of the owner trickles down to the rest of the team. If the owner is a fair, easygoing, solution-oriented party with common sense, the project normally flows easier, and the rest of the team members are more relaxed and better focused on the project.

A tough owner who is fast to blame and gets on the case of his/her construction manager or architect leads to a project where most team members spend a lot of time worrying about protecting themselves in addition to getting the work done.

To add some fun to the process and since we are talking about the anatomy of construction projects, I developed an easy to remember set of key steps for successful project Management. I called it the ANATOMY formula:

Analyze the project's elements and potential risks before starting the project

Network with the team members and partner with them before starting the project

Activate your project logs, initial budget sheets, and read the documents from A to Z

Transfer your research comments and observations to your project's team

Organize the procedures to be followed on the project

Make sure every team member has a clear description of their roles and responsibilities

You are the leader who will set the tone for the project; keep it positive

The rest of the book will address the above concepts in their relevant chapters in detail.

Chapter 4
Construction Manager

Construction management is a fundamental realignment of traditional construction relationships. It removes management functions from both the contractor and the designer and places it in a specific entity: The Construction Manager.

Think about it; if the architect is managing the project, there is a bias there as the architect tends to want to defend his/her design and argue against claims resulting from errors and omissions. The contractor can't play that role for the obvious conflict of interest. The construction manager, theoretically, is outside this interest group that keeps all in check and should be a fair arbiter of the conflicts and issues.

Who Hires a Construction Manager?

Typically, construction managers are hired by public entities with a large capital improvement program, like school districts, cities, and counties. Some large private-sector developers hire construction managers as well.

Contractors usually manage their projects using in-house project managers. So in this chapter, we will focus on the construction manager working as a consultant/owner's representative.

Why Do Public Entities Hire Consultants?
- The program is too large to handle with their own staff.
- Their staff doesn't have the needed experience.
- It is safer to allocate the liability to others.
- Replacing an unwanted consultant is easier than replacing an employee.
- The CM firm is open to the risk of costly litigation or damaging reputation in the market.

Construction Management Practice

There are several forms, and they all fall into one of these two categories:

Agency
  • The agency CM acts on the owner's behalf, advises on, or manages the project.
  • Does not perform design or actual construction work.
  • In this format, the CM is promising his/her best effort and has to follow the industry's standard of care for this kind of service.
  • The CM can still be held liable for negligent acts.

At Risk
  In this format, the CM's role includes some form of guaranteeing a result, like the final cost not exceeding a certain amount. The fees for this service are usually higher due to the higher risk.
  The world of construction management is very wide and can include numerous and various type of tasks. Services offered can include any one of the following:

  • Program Management. For example, a school district just got a $200 million program and needs the CM to oversee the following:

  1. Determine the set of projects that will be done
  2. Develop an execution management plans of how this program could be run
  3. Develop a master budget allocation and a master schedule
  4. Reports
  5. Etc.

  • Management for one or more specific projects in the program:

  1. Develop the scope
  2. Manage the design phase
  3. Advertise the project
  4. Manage the contractor up to completion of work
  5. Refer to chapters 5, 6, 8, and 9 for a more detailed discussion about the above phases

CM Services
  What should you do when you first get hired as a CM/owner's representative? The first step is to understand your client. After

all, you are there to help the owner meet his or her goals. So set up an initial meeting and ask about the following:

- What are the goals of this new program? What would you like to build?
- Any promises made to the community
- Timeline
- Funding restrictions
- Any politically sensitive issues
- Owner's internal processes
- Owner's team structure and the best way to interact with the CM team
- Expected deliverables
- Budget allocations
- Community involvement
1. Community oversight committee
2. Political considerations
3. Elected officials influence
4. Be aware of the commitments
5. Be aware of the sensitive issues
6. Promises made
7. Always work through your client to interact with the community representatives
- End users of the projects
1. The occupants of the various projects
2. Understand their functionality
3. Their dynamics
4. Understand their project objectives
5. The use of the building
6. Political connections

What is expected of you as an owner's representative? The same thing you expect when you hire an accountant or go see your doctor. You expect them to know all there is to know about your problem and how to solve it. You expect them to have the knowledge and experience to deal with your issue. The same applies to the owner. He/she hired you as the expert in construction. You are expected to:

- have the proper education, right background, and expe-

rience for their project
  • know how to manage the project
  • know how to manage designers and contractors
  • be organized enough to keep track of issues and meet timelines and budgets

Real-Life Gauge

As you are hired to manage several parties involved in the project, one of them is your client, the owner. That may be your toughest challenge as the owner is the one who hired you and is paying the bills. The owner sets the tone of a project. If the owner is flexible, easygoing, fair, and supportive, that is reflected in your work and gets transferred down to the rest of the team.

The key to your success is to build good rapport with the owner, your client, and get him/her to support you as you face the multitude of decisions and problems. The owner may be a small outfit or a huge bureaucracy, which can complicate the simplest issue. You may find yourself in a position in which you know what needs to be done but can't do it because of some direction or policy of the owner.

You really have to understand the dynamic within the owner's structure beyond your immediate contact that you deal with every day. Because that person has to report to some higher-level management body, maybe he or she is controlled by another set of deadlines and commitments that you don't know about. Those may be the driver of his/her decisions that you get.

So it is not always about you and what you think needs to be done.

Here is an example. I was in the position of an owner's representative when I made a comment about the program that scheduled underground electric conduits to be placed after the areas get new asphalt pavement. I raised the obvious question, which was "Why don't we place those conduits before we place asphalt?" You would think that was a simple question. However, it took months of discussion and managed to rock several people's boats so that finally, we gave up on this quest and concluded that it was better to just pave now and excavate later.

I can also give you examples about other owners where the decisions were so easily taken and issues flowed much simpler.

The best solution to this is to establish a strong and close relationship with your owner's contact person. You have to make him/her trust that you are there to look after his/her best interests so that person shares with you those unknown factors, and together you can discuss the best solutions that work with that organization.

Best Practices

- Build a good relationship with the owner.
- Write down the goals, especially the sensitive issues.
- Develop a method on how to track issues.
- Regular checking of logs and schedules.
- Meet regularly with the owner.
- Continuous communication with the owner.
- Keep the owner informed about potential problems.
- Take the steps that get the job done *and* keep the owner out of trouble.

Chapter 5
Planning Phase

A construction project goes through the following typical phases:

- Planning phase
- Design phase
- Bid/Award phase
- Construction phase
- Closeout phase

This chapter will introduce you to the planning phase while we address the others in following chapters.

Let me start the discussion from the end. I'm sure you are familiar with the fact that several projects get into claims. Of course, there are many reasons for the claims, but the main source of disputes is some kind of a problem in the contract documents. That means claims avoidance starts from the earliest stages of preparing these contract documents; that is the planning phase.

A lot of owners rush through the preconstruction phase to get into construction. They end up paying through the roof   in change orders and claims.

One of the main responsibilities behind hiring a professional like you, the construction manager, is risk management.

- Risk management is simply watching out for potential liabilities against the owner, like change orders and claims.
- Risk management starts on day one, not when you hit the problem.
- Risk management starts from planning phase, and that effort continues all the way to completion of work.
- The construction manager managing these various phases of the projects should review documents with an eye for claims.
- In order for you to properly manage risk, as a minimum, you need to be fully aware of the contract documents and the owner's

and contractor's responsibilities.

This planning work can be at a program level or a specific project level. What is the difference?

Program Level
Let's take a school district, for example, that just passed a bond for $400 million to improve its existing school and build new ones. At this level, you are involved in program management.

Project Level
Take the abovementioned $400 million program; this planning is dealing with each specific project in the program.

Program-Level Planning
Let's stay with the above program as an example. At this level, your client got $400 million and needs your assistance in figuring out what to build and how to manage and execute such a large program.

The owner is going to ask you to figure out how much this is going to cost, how long it will take to get done, how much of the available funds will go to overhead versus construction, etc. Now, different program managers may come up with different organization models. To make this point clearer, here is one example model:

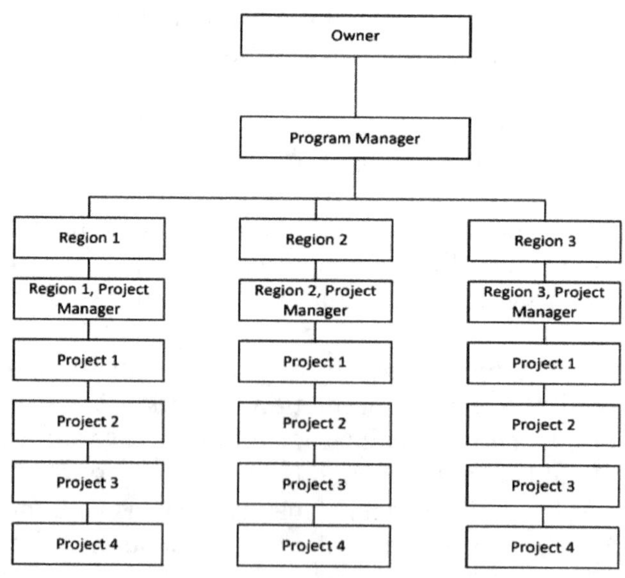

So let's say the above model is the one you came up with. You will have to figure out the following:

- Based on the master schedule and the duration of this program, how much is the cost for the staff serving the above model?
- How many architects are we going to hire? Because that is also a cost issue.
- How will the reporting be between the regions and the program manager? Will there be weekly meetings, for example?
- What kind of forms are needed for reporting and tracking?
- How many projects in each region, and how much is the estimated budget for each?
- How many construction managers should we hire for each region?
- Develop and executive management plan that defines the procedures for the whole team.
- Etc.

Before you do all that, meet with the owner and ask about the following:

- Program goals
- Program funding
- Political considerations
- Environmental issues
- Develop an executive program management plan
- Develop a master budget allocation plan
- Develop a master schedule

Let's go through each of the above tasks.

Program Goals

This may be a program that passed the last November election based on promises made to build new schools and remodel the existing ones. There may be specific promises made, like all school playgrounds will receive new asphalt, etc.

This will be important for you to know when it is time to figure out the projects that will be built in each region.

Program Funding

Do we have all the $400 million at hand now, or are we getting it $50 million at a time every two years? How will that impact the program plan?

We look at the cost line items; there are the construction costs and the overhead costs. Once you develop a conceptual estimate for each of the projects and plot the timeline of expenses, you can decide which projects can be done for every two years until the next $50 million comes in.

Another consideration may be that funding comes in with time constraints. For example, the state won't release the next $50 million until 80 percent of the first $50 million were spent.

Another consideration could be that access to funds requires certain procedures that take time.

There may be more or less considerations, but the point is that knowing these facts will impact the program management plan.

Political Considerations

Ask the owner about any political issues that you need to be sensitive about. For example, if there are some historic buildings on some school campuses that should be left as is, or if the public in the community of Region 3 are very adamant about not cutting trees, or if some politician promised that the public will be consulted before selecting the exterior paint colors of the schools, etc.

Again, these will impact your proposed executive management plan and the guidance that has to trickle down to the region's project managers and from there to the various architects and projects.

Environmental Issues

The big issue when it comes to remodeling existing schools is the presence of asbestos and lead. What you will need to figure out, for example, is who will do the testing? Will it be done in-house, or should we hire consultants? What kind of testing and inspection will be implemented? Etc.

All this impacts the cost and timeline of the program.

There may be requirements to do an Environmental Impact Report (EIR) before building any new school. That will impact cost and time.

Develop an executive program management plan. Considering the above issues and probably more depending on the program you will have to develop an executive program management plan to outline how the program will be organized, procedures, forms, cost control, document control, meetings, reporting, etc.

You may be asked to present this to the owner, school board, city council, etc. Not to divert from the topic of this book, but I think it is worth your time and money to invest in learning public speaking skills. Just a thought!

Project-Level Planning

Based on the above model, *you*, the program manager, developed a model to manage this $400 million program. As per the above model, each region will have various projects.

In this segment, we'll be talking about what is involved in managing each of the specific projects.

Again, first thing you have to do is meet with your client, the owner, and understand, in general, what the project is all about. What does the owner want to achieve? For example, the community has been asking for a new library that has spaces for career training or private tutoring or a community center that the local residents can visit for youth consulting services. Ask about budget, funding, promises made, timeline, milestones, etc.

If you understand those factors, then you can execute the planning phase more efficiently. Let's go through these concepts in more details.

Meeting the Stakeholders

Who are the stakeholders? This can be two people or a whole set of groups that can include the following:

- The owner
- The end user of the facility
- Local community groups
- Oversight committee
- Providers of funding
- Politicians
- Media

What should you find out from the stakeholders so that you can develop the best approach to execute the planning work?

- What type of facility they want to build
- Budget
- Funding
- Timeline and milestones
- Promises made to the community
- Political considerations
- Who is the end user of the facility
- Scope restrictions

Were there any steps already taken for the project?

- Talk to the end user. That is the group that will occupy the building.
- Find out what their needs are for the new improvement. If you don't, then the architect may include a new gym in the design that this school never wanted in the first place.
- Understand their dynamic. You'll find out that the architect should design John's office to be close to Jennet's because these two have to interact on a daily basis.
- Do not make promises you can't meet considering program restrictions we talked about above.

Visit the Site

- Understand the site, its restrictions, and other existing conditions that need to be addressed.
- Take the architect with you as there may be some code-regulated issues you may not be aware of.

Organize Scoping Meeting with the End User

This is when you meet with the end user and other stakeholders to discuss the scope of the project. You need to hear what they need and discuss what the program can and can't deliver.

These meetings, if not organized and well planned, can go for months without any result. So when you start these meetings, define the objective and timeline of the discussion, making it clear that after a certain date, it is safe to proceed with the design.

Planning Phase: Schedule/Phasing

Part of the planning for the project is figuring out the schedule of the project, defining important milestones and phasing.

Here is an example. Let's say the project is remodeling an existing school. Unless the plan is to vacate the school completely, the construction will have to be done in phases. That will require planning of what these phases are, how long each phase will take, where to relocate the students and staff during each phase, if we should bring temporary buildings for the relocation, what type of power/data phone arrangements are needed for the relocated group, etc.

If you don't think about these issues and you advertise the project for bidders without phasing specified in the bid documents, the future contractor can argue that they did not bid the project in phases, since phasing was not specified. That will result in a big change order.

There goes the budget, and your client, the owner, will not be happy with you as you were hired based on the fact that you are a professional who knows what you are doing.

That strengthens my previous point that risk control starts from day one.

Planning Phase: Procurement Method

We will go over this in details in a later chapter. How will you bid this project? These are the choices you have:

- D-B-B (Design-Bid-Build)
- DB (Design-Build)
- CM (Construction Management) – Multi-prime
- Lease Lease-back

Each one of these procurements methods is managed differently and may impact your work during the planning phase.

Planning Phase: Budget

Cost control is one of your main tasks as a construction manager. Again, this control starts from day one.

As you are developing the scope in this planning phase, you need to make sure that the potential cost of the project does not go beyond the set budget. Please take note of the following tips:

• As the scope is developing, run a conceptual estimate. Once the numbers get close to the set budget, raise a red flag and readdress the scope.
  • Be aware of allocated funds for your budget.
  • Be aware of constraints as to what can't be spent on.

Planning Phase: Funding

Earlier we talked about funding issues for the whole program. It is important for you to know about the funding relevant to your specific project as it may also impact the planning phase.

  • Where is the money coming from?
  • What is the timeline for the funding?
  • Any constraints to get the funds?
  • What is the client's process of allocating funding to the project?

Exercise

If you have time, go ahead to pretend you are planning a remodeling project, let's say your local public library. Do the following:

  • Write the scope of work
  • Estimate the budget
  • Describe the phasing and schedule
  • Determine what form of bidding you want: D-B-B, DB, or CM/Multi-prime, etc.
  • Present it to someone; let them challenge you

Real-Life Gauge

As mentioned in several chapters of this book, your main challenge to performing the construction management task is people. Human nature will be your biggest challenge. Remember to maintain an excellent relationship with your immediate contact from the owner's team. As you see from the above, you will be dealing with so many people with various agendas and problems. If the owner, who is the main party that hires everyone else, has your back, then your job will be so much better.

To do that, do what you need to do to get the owner to trust you. The owner should trust your experience to be what is needed for the owner's program. Maintain regular communication with your client. That could be daily. Maybe meet over lunch away from the workplace and talk about the program. The owner has other people he/she has to report to. Make sure the owner always looks good in front of these groups.

Other than the owner, be the facilitator of issues for the rest of the key players. Be the person who brings in good news and solutions. In short, if you help others win, you succeed.

Best Practices

- Build logs to track issues.
- Build schedules for the various tasks to meet deadlines.
- *Keep these logs and schedules visible to you.* A schedule or log that you cannot see on a daily basis is useless. Things will fall through the cracks.
- The walls in your office should be full of schedules and logs. These should be updated regularly as things change. Simply walk to that wall and redline updates. If you can prefer to do that electronically, so be it, as long as these logs and schedule are visible on a daily basis.
- Develop a system to track issues.
- Update logs and then do the task.
- If you get a request or think of something to do, write it down in a to-do list that is also visible on a daily basis.
- Finally, be the pleasant person who all parties feel they want to call for solutions.
- Start your work with a partnering session with key players. It breaks the ice.

When you are organizing meetings, whether it is with the end user or the design team, make sure to do the following:

- Lay out the plan and agenda for the meeting in advance.
- Put timelines to the discussions and conclusions.
- Avoid redundant meetings.
- Don't invite people who have nothing to do with the discussion, or they will feel that their time is being wasted and will bring down the energy of the meeting.
- Have sign-in sheets to document who was there.
- Take notes and distribute meeting minutes to the attendees.

Chapter 6
Design Phase

Now that you have completed the planning phase, you have a good idea about the scope of the project. It is time to start the design phase.

Before we get into those details, let us talk a bit about hiring an architect. To start with, there is a difference in the hiring procedures if the project was a private sector project or a public works project.

In the private sector, the owner can hire any architect they want and for the fee that they want. However, it is a totally different story in the public sector. For a public entity like a school district, a city, or a county to hire an architect or other professional service like construction management, they have to advertise a "Request for Proposal" or a "Request for Qualification". The idea is that this is public money and a public entity, so there should be no preferential treatment for one design firm, and a fair chance to compete needs to be provided to the public.

Here is how it works:

• The public entity prepares and publishes a request for a proposal identifying all the details of the project, including the scope, the needed service, duration, and other details of the project.

• Typically, the format of the proposal is identified in the request.

• The method of scoring and evaluating the submitted packages. This is an important step to demonstrate transparency in how the owner will evaluate bidders. That serves the objective of being fair in considering firms.

• There may be an interview involved, which will be scored per the declared procedure for qualifying the firms.

As mentioned before, risk management starts from day one. The preconstruction phase is the time to foresee all potential problems and take steps to avoid them. Selecting the right designer is obviously a big step in that direction. Therefore, qualifying architects and choosing the right firm is important to produce good quality design documents that prevent the main source for future claims.

Here are some issues to examine while qualifying architects:

• Check the type of projects they design to decide if they are a good fit for your project.

• Get some references.

• Visit some of the completed projects that they designed to check the quality of their work.

• Visit their office. That will show you what kind of an operation they have, their staff, and you can look at some of their designs.

• Ask about the subconsultant engineers that they will use on the project. Those are the various engineers. For example, if they have established relationships with them. That is important since the architect is the main coordinator among all design disciplines.

The architect is the main design professional on a project. There are several trade disciplines that have to be designed like electrical, plumbing, structural, landscape, etc. These specialty engineers are typically hired by the architect, who has to coordinate between these trades to make sure the plans are complete and without conflicts. Since each of these engineers has their own firm, they are contracted by the architect as subconsultants.

Here is a typical list of subconsultants:

• Structural Engineer - Designs the structure of the building.

• Civil Engineer - Designs the site work like grading and site concrete.

• Electrical Engineer - Designs Electrical

• Mechanical Engineer - Designs Heating, Ventilation and Air Conditioning (HVAC) and plumbing.

• Interior Designer - Designs the interior spaces of the building.

• Landscape/Irrigation - Designs the layout of the landscape and the relevant irrigation.

• Soil Engineer - Tests the soil for structural purposes and how to prepare it for the new structure.

• Environmental Consultant - Tests the areas for hazardous material and specifies the measure to remove them.

How Do Architects Price Their Fee?

We discussed above the planning phase. You can hire the architect at the planning phase where the scope is being developed or after the conceptual design scope has been developed. Sometimes the owner's budget dictates if an architect can be hired early on or not.

Whatever you decide, the owner approaches the architect with an idea about the project and a set of required design services.

The architect's fee is usually based on the anticipated number of hours plus other costs like printing or copying and travel time. The same applies to the architect's subconsultants.

Each of these firms estimates their hours and costs. The architect reconciles all the data into one fee proposal to the owner.

That is why as a construction manager, it is important that you respect the architect's time and not waste it in unnecessary meetings or unlimited changes as this eats up the architect's budget, especially before the construction phase where you will need the architect to support that phase.

Remember that in order for you to succeed as a construction manager, you need to make sure that all parties are winners; after all, they are the ones who are taking care of your project, whether it is the architect, the contractor, the inspector, etc. As a construction manager, you have to understand where each party comes from to know how to best deal with them. So for architects, time is gold.

How to Best Deal with Architects?

From the above paragraph, we know that respecting their time is important. Here are some tips.

1. If you hired the architect after the planning phase, this means that the time of the architect is set to design the desired scope. So, approach the architect with a clear scope and as much parameters and information as they need to proceed with the design. Don't waste their time because of your lack of organization and planning. The result will be the depletion of the architect's budget from an early stage and subsequently a tense architect and a poorer quality work product.

2. Set clear timelines for the various design deliverables from the architect.

3. Follow up with the architect regularly to check up on the status of the design. Remember that your project is not necessarily everybody's priority.

4. Be part of the team. If you are the owner's representative, the architect will need your decisions and will need your input to get direction on how to move forward. You should be perceived as a person who solves problems.

5. Don not make too many changes. Some owners don't understand the impact of changes in the middle of the design stage. For example, a simple request to move a door from one place to another may mean the alteration of fifty details between floor plans, sections, and details. So try to finalize all scope issues before the start of design. The best way to achieve that is to manage the owner's group during the scope development and make it clear to them that after a certain date, you will not be able to make any more changes.

6. Conduct plan reviews periodically and run estimates for each deliverable before moving forward with the design. This way you avoid exceeding the budget and control scope creep.

So How Does the Design Phase Work?

Start the design effort by a kickoff meeting with the architect, their subconsultants, the owner, and other stakeholders. Discuss the scope of the project, the timeline, the required deliverables, and other related matters. Organize this meeting to lay out *all* the elements that the architect needs for the design, so once they proceed, we minimize changes and interruptions.

We mentioned the term *deliverables* a lot in the above paragraphs. These could be different in different contracts. To explain the point, let's consider the following set of deliverables:

- Conceptual Plans
- 30% construction documents
- 60% construction documents
- 90% construction documents
- 100% CDs and specifications

Conceptual Plans/Site Verification

At this point, the architect prepares plans showing the

concept of the design and visits the site to incorporate site conditions or raise some limitations due to the site condition.

- Needed Action

a.   Review these documents with the end user to get their sign-off on the scope. The presence of the designer can be valuable for the discussion.

b.   Verify that the conceptual estimate is within your budget. If not, address the scope changes.

c.   If the scope and budget are acceptable, instruct the designer to proceed to prepare the 30% documents.

30% Design

Now the designer has more details and information on the plans. However, since they are still at 30%, your review at this stage is a cursory review.

- Needed Action

a.   Take a cursory review of the plans.

b.   Visit the site to look at site conditions that need to be addressed.

c.   Verify that the designer did not deviate from the accepted scope.

d.   Get a revised cost estimate to check against your budget.

e.   Give your review comments to the designer and keep a copy to yourself so that you can verify that the comments were incorporated in the next submittal.

60% Design

At this stage, it is important to have a more detailed review to avoid having the designer move on to the final phase, and then ask them to do all kinds of changes or revision. Designers lose money this way, and it will affect the quality of their service. It is similar to an inspector waiting until the contractor completed his wrong installation, and then requesting the work to be removed in lieu of making a comment early on.

- Action

a.   The main purpose of a 60% review is to verify that the scope is completely indicated on the plans and all site conditions have been addressed.

    b. Review the revised cost estimate to make sure you are still within budget.

    c. I recommend getting an independent estimate in addition to the designer's estimate to double-check that you are still within budget.

    d. Complete your comments and meet with the designer to discuss with them before you give the designer your final set of comments.

    e. Please remember to keep a copy of your comments to make sure that your comments have been incorporated in the 90% set of plans.

90% Design

Assuming you have no further comments, the 90% construction documents should be at a stage ready to be advertised. At this stage, you need to do a very thorough review of the documents.

- Action

    a. At this stage, you need to read the documents from A to Z. You are looking for errors and omissions, checking discrepancies, and looking for change orders and claims.

    b. Verify that your previous comments have been incorporated.

    c. Read all the general notes on the plan.

    d. Visit the site to verify site conditions.

    e. Review the specs for a complete scope of work description.

    f. Make sure a phasing plan is developed. In some cases, a plan showing the phasing is needed.

A common mistake that owners do is rush into construction, only to pay through the nose in change orders and get into claims.

Think about it: What is the source of most of the disputes? Discrepancies in the construction documents, errors, omissions, and unaddressed site conditions. That is why it is worth every penny to allocate a budget for some plan reviews before the documents get advertised and get into the construction phase.

Two types of review are very important; constructability review and value engineering review.

Constructability Review

I have conducted dozens of this type of reviews. Basically, I put on a contractor's hat, looking for change orders and items that can turn into claims. Just to give you an idea, look at these few examples, that will make the point. These are actual examples that I came across:

Example 1. The plans show a ramp to comply with ADA requirements. Well, I take that plan to the site to check things out, and there is a tree right where the plans show the ramp. Hello, change orders.

Example 2. A general note says "All existing air-conditioning louvers to be repaired as required." An immediate red flag for a claim. The scope, as described in this note, is so vague that bidders have no way to quantify and estimate this scope. That is a potential claim where the contractor claims this is an extra, and an architect will argue that the note is in the plans and, therefore, in the scope.

Example 3. The plans show a slab on grade with no indication of the slab's thickness or the size of the rebar.

Don't be surprised; I have seen worse. That is why this review is very important before we get into the construction phase. If you catch these issues early and add the needed details to the plans, then they will be priced competitively versus the higher costs in change orders.

During this review, you are looking for the following items:

- Buildability of the project following the plans.
- Missing details and information.
- Conflict between various documents.
- Verify that site conditions have been addressed.
- Look for potential claims and change orders.

This is *not* a design review. Remember, the architect is the licensed professional who is legally responsible for the design. Don't make the mistake of stepping into that area of liability by dictating design details to the architect. It is better that you suggest them to the architect and let him/her make the changes. The last thing you as a construction manager/owner's representative wants in case of a leak or a failure is for the designer   to say, "This detail was the owner's, not mine."

How to Best Conduct a Constructability Review?

I have encountered many approaches to conducting this review. The main challenges are the following:

- Efficient method of documenting the comments.
- Effectively communicating the comments to the architect.
- The interaction between the reviewer and the architect.
- Getting the revised comments back and making sure that all five hundred comments were actually incorporated.
- Setting a time limit to complete the review.
- Budgeting control regarding the cost of conducting this review.

Here are some of the methods that I have seen:

- Method 1:
  a.   Conduct the review and insert the comments in a spreadsheet. See Contractor's Comments example in the end of this chapter.
  b.   Email the comments to the architect.
  c.   The architect works on the comments and sends back the responses. See Designer's Reponses example in the end of this chapter.
  d.   Receive the architect's responses with the revised plans.
  e.   The reviewer confirms that the comments have been incorporated and the plans are ready to be released to advertise.
  f.   The process is repeated until all comments are taken care of.

- Method 2:
  a.   Review the plans and redline the comments on the plans next to the issue in question.
  b.   Meet with the architect/engineers and go over the comments.
  c.   Give the redline to the architect.
  d.   Architect incorporates the changes and revises the plans.
  e.   The reviewer verifies.

I have tried both ways and prefer method 2 by far. Here is why:

- Method 1:

    a.   The architect gets a list of five hundred comments, let's say. Some comments he/she will understand and some he/she won't.

    b.   You end up going back and forth to explain what you mean.

    c.   You may get a written response from the architect with more questions than answers or wrong answers. Say hello to another 150 emails back and forth.

- Method 2:

When I went through my redline comments face-to-face with the architect with the plans in front of us,

    a.   We had a more educated discussion.

    b.   some comments were simply voided after the discussion.

    c.   We were able to brainstorm and exchange opinions faster than dozens of emails back and forth.

    d.   When the architect takes the redline plans to work with, he/she knows exactly what the issues were.

Sometimes you may have a client who is dictating a whole different approach. Well, you decide what works better for your project. The main goal is to foresee all potential problems before the project gets to the bidding stage.

See below an example of a spreadsheet and a process that you may use.

Value Engineering Review

This is a different type of review. The constructability review was more a buildability review. The value engineering review is basically looking for other options that cost less but achieve the intent of the scope and the design.

For example, you may find that we can use less HVAC units than what is shown on the plans, using different sizes and still providing the needed airflow. Or you may suggest less costly light fixtures and provide the same degree of lighting. Or you may look at a construction detail and suggest a different way of building it, which will cost less, etc.

You get the picture. For this type of a review, it will be a good idea to hire specialty subs and engineers to help. As far as the process goes, you can follow that same suggestions outlined above for the constructability review.

Contractor's Comments											Designer's Response			
Spec Section	Plan #	Detail #	Location on Plan	Comment	Reviewer's Name	Company	Trade	Date of Comment	Item ID #	Cost Impact	Response to Contractor's Comment (Issue response after a discussion with Contractor to maintain overall schedule)	Reviewer's Name	Date of Response	Revised Plan or Detail No. (Write Item ID # on new plan or detail)
Insert a Spec. Section if your comment is relevant to information noted in a spec section	Insert a plan # if your comment is relevant to information noted on a plan sheet	Insert the detail number on the plan sheet where the comment was noted	If the comment is from a plan sheet and not from a specific detail, describe where on the plan are you referring to.	Describe in details your comment. Include a suggested solution if possible. If you issued a sketch or any other back up document, please indicate the Item ID # on each page. If you wish to send back a follow up comment regarding the same issue, please insert a row below the original comment keeping the same Item ID #, but a different date of course. This will help keep track of the development of the same issue	Insert your name	Insert your company's name	Insert your trade, Concrete HVAC etc	Insert date of your comment	If your name is John smith and this is your comment number 1, then your comment's ID # is JS - 1C. Next is JS - 2C etc…. The C at the end indicates that the item was listed on the Constructability Review sheet versus the Value Engineering Sheet. Each person to maintain his/her sequence since comments may come in from various sources. In case of similar initials, Hilbers will adjust when they reconcile all the spreadsheets into one.	Insert the net cost impact of the items, That is an add of $x or a saving of $x	Insert your response to the comment. For the sake of time efficiency and overall schedule, please insert your comment AFTER YOU HAD A DISCUSSION WITH THE CONTRACTOR. This will reduce back and forth paper work and will avoid a tedious process to reconcile comments.	Insert your name	Insert the date of your response	If you issue a new detail or a revised plan as a response to a comment or several comments, please indicate on the each detail or plan the Item ID # or #'s so that we can track that all comments have been addressed

Chapter 7
Various Project Delivery Methods

Most projects follow the standard process of what is called Design-Bid-Build (D-B-B), which is this:

- Architect/engineers *design* the project.
- The owner advertises the project for contractors to *bid*.
- The successful contractor *builds* the project.

However, some projects may be better managed and advertised following a different delivery method. The owner may turn to you, the construction manager, to advise as to which method is best for the project.

In this chapter, we will explore the various methods and discuss the principles and the challenges of each. We will explore the following:

- Design-Bid-Build (D-B-B)
- Design-Build (DB)
- Integrated Project Delivery Method (IPDM)
- CM – Multi-prime
- CM at risk
- Guaranteed Max

Design-Bid-Build (D-B-B)
This is the most common delivery method. It is basically the way it reads; the project is designed, then advertised to bid, then a contractor builds the job; Design-Bid-Build.

In private-sector jobs, the owner has much more flexibility regarding which architect and contractor to hire. After all, it is the owner's money. In public works, it is a whole different set of rules. We will cover this in more details in Chapter 8. The hierarchy structure and contracting structure of D-B-B is represented by the following diagram.

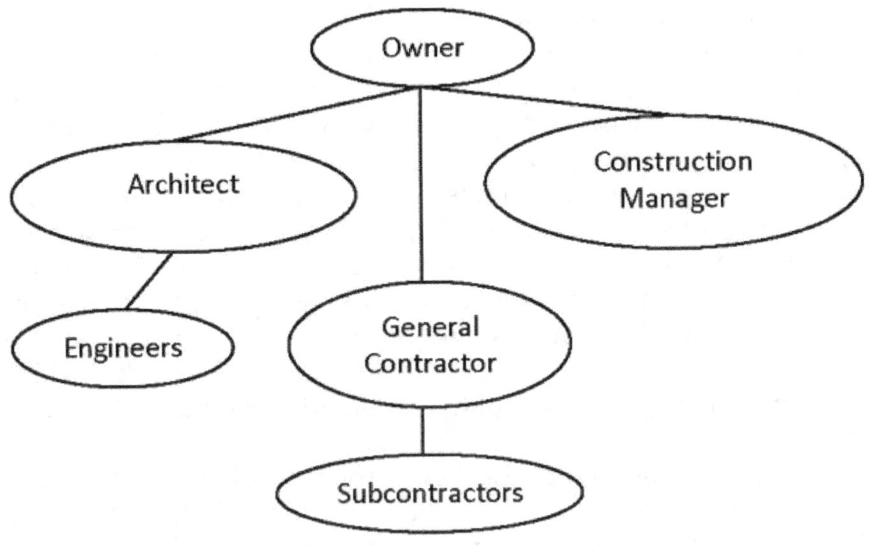

Design – Bid-Build

In public works projects, the bids are submitted in closed envelopes, and the lowest responsible/responsive bidder gets the project. We will go over that in Chapter 8. What I want to explain here are the perspectives of the people who like this method and the people who don't.

Those who like this method argue that this approach provides more opportunity for contractors to bid on projects at an equal level with other larger firms.

Those who don't like this method, mostly public owners, feel that they will be stuck with a contractor that they did not choose, only because that contractor submitted the lowest number. There is some truth to that, as a public owner may get a contractor without much experience or financial capability to execute the project.

I remember a small project ($350,000) that I managed representing a public owner. The contractor did not have public works experience. That project took more of my time and had more problems than another two $7,000,000 projects combined.

As a result, public owners are publishing pre-qualification packages to a list of contractors that they feel will be qualified for their projects. Here, we get into a different set of problems that contractors complain about. Here are some examples:

• The questions in the pre-qualifications eliminate small contractors where only large contractors can qualify, like "List five projects in the past five years above $8 million that included a public facility with a pool." So if there is a contractor who completed three projects with pools, one for $20 million and two for $6 million, this contractor won't qualify when they are perfectly capable to do the $8 million projects. Besides, general contractors will hire specialized pool subcontractors who probably have done over twenty pool projects. Owners argue this is fair; contractors argue that this is against the spirit of public works, which is supposed to give everybody an equal opportunity to compete. You decide!

• Another question: "List five projects in the past five years above $8 million that included a public facility with a pool." So if there is a contractor who completed three projects with pools, one for $20 million and two for $6 million, this contractor will not qualify when they are perfectly capable of doing the $8 million projects. Besides, general contractors will hire specialized pool subcontractors who probably have done over twenty pool projects. This approach opens room for owner panels scoring the applications to have favoritism toward some contractors. After all, scoring in some cases are purely subjective.

The other problem with D-B-B is the disconnect between the designer and the builder. That is true because the architect designs the project long before a contractor enters the picture. When we get into the construction phase, the contractor starts pointing out all the mistakes and missing information in the design, or they point out that a specified product is discontinued, etc. That means change orders and arguments. This leads me to talk about the next delivery method, the Design-Build method.

Design-Build

The idea here is to bring the builder and designer together from the start of the project so the generated construction documents incorporate input from designers and builders. This will reduce change orders and claims of errors and omissions. Think about it: using the D-B-B method, all errors and omissions are the architect's fault. In the D-B model, the builder and the architect prepared the plans together, so these errors and omissions are both of their mistakes.

The intent of this chapter is to introduce you to this concept since covering all aspects for Design-Build is a whole book. You can look up many books on the topic, and the website of the Design/Build Institute is another great resource (www.dbia.org).

As you can see from the diagram below, the owner hires a contractor (most of the time) as the prime contract. The contractor hires an architect, and together they become the "design builder." The architect hires their engineers; the contractor hires their subcontractors. This whole team will collaborate to develop the design and the construction documents.

I am sure you can see the benefit of the designers and builders going over details and material together. Imagine the benefit of an electrical subcontractor talking to the electrical engineer from day one instead of getting into construction and then bringing up everything that the plans missed. The benefit to the owner is a more efficient design, lower cost, faster completion, and lesser claims between builders and designers.

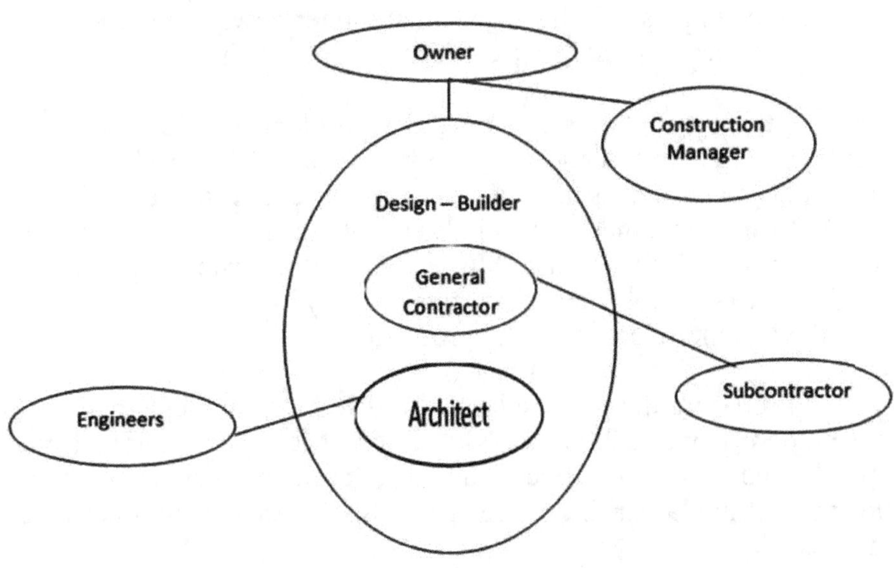

Design- Build

Here is briefly how it works:

• The owner develops the scope and design criteria, some-times with the help of an architect, but not the same architect in the Design-Build team.

• The criteria is advertised for Design-Build teams to submit proposals and compete with other Design-Build teams.

• The successful team collaborates and meets regularly to de-velop the design.

• The contract defines design deliverables to review with the owner before proceeding. At that time, the owner verifies that the estimate and scope are still within the original criteria.

• Construction starts after the final construction documents are approved by the owner to proceed.

Obviously, the construction management process is completely different from managing the D-B-B process. From a career standpoint, you, the construction manager, can fill many positions here. You can be the following:

• The owner's consultant who helps in preparing the initial cri-teria and the request for the proposal and interviews the bidders.

• The Design-Build project manager hired by the design-builder.

As the concept makes sense, the problem is in the procure-ment method. I have seen Request For Proposal (RFP) packages that are twenty pages and some that are large five-inch binders. Unlike the D-B-B method, the lowest bidder is not the only criteria to be the successful firm. The owner publishes the scoring criteria, giving the public owners a vast leeway to require any set of qualifications that they want. Here are the problems:

• Preparing the proposal per the listed requirements can cost the bidders from $30,000 to $1 million. Many contractors can't afford to pay that out of pocket and not get the project. That leaves this market the larger firms. Preparing this package sometimes take months.

• Again, the questions in the RFP may eliminate a lot of firms that can, in fact, perform the work. For example, "List three De-sign-Build projects that you worked with the proposed architect on projects of similar size and scope." Give me a break. So if you are experienced and have built similar projects, and you teamed up with another experienced architect who designed similar projects, you will not qualify for this proposal.

I know so many contractors who won't touch a Design-Build bid because of the time and cost it takes to prepare the proposal and because of the feeling that they don't stand a chance against larger firms that have Design-Build projects in their portfolio.

This is where contractors feel that D-B-B gives a fairer opportunity to get projects. I have often voiced this issue at seminars, hoping someone figures out a way to make the Design-Build procurement more inviting to smaller firms because the concept makes a lot of sense.

CM/Multi-Prime

Well, here is the idea behind this. When general contractors bid a project, they get prices from various subcontractors and add their markup on top of the subcontractor's markup. Owners thought, "Why should we pay this much? If we hire the various subcontractors directly, we can save the general contractor's markup."

The definition of a prime contract is a contract directly with the owner. So since the owner will contract directly with the various subcontractors, this forms the "multi-prime" portion of this method. Having a construction manager handle the project makes the name "CM/Multi-Prime".

The diagram below demonstrates that relationship.

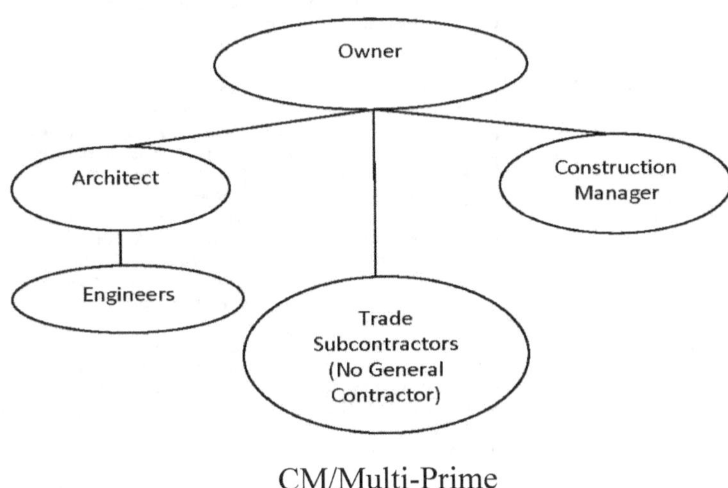

CM/Multi-Prime

As the idea seems simple, however, the procurement and the management of this method is much more complicated than the standard D-B-B procurement. Please note the following table.

As you can see, your job as the owner's construction manager gets more complicated with this delivery method. So why use this method? Sometimes, the owner is sold on this idea, and that's why they hired you, the construction manager, to worry about the details. Or for some project types, this method sounds like the most suitable. You decide!

Task	D-B-B	CM-Multi-Prime
Pre-Construction Phase		
Design Process	Architect/Engineers design and prepare one set of Construction Documents	Architect may have one set or multiple sets for each bid package
Preparing the Bid Package	Prepare one set for the General Contractors to bid on	Prepare multiple bid packages for each prime contract. Packages can be include one trade, like Plumbing or a set of trades, like Bldg A could be one of the packaged
Advertizing the Project	Advertize for one bid package	Advertize for multiple packages or separate advertisement for each package
Bid/Award process	Receive, evaluate and award one package	receive, evaluate and award multiple bids for the various packages
Awarding Prime Contracts	One award to the General Contractor	Multiple awards for each package
Construction Phase		
Coordination on site	The General Contractor coordinates with all Subcontractors. In this case the Construction Manager only communicates with the General Contractor who handles the rest of the trades	That is a challenge as the CM has to coordinate among all the trades and there is no one Contractors who is in-charge and who can held liable for lack of coordination.
Requests for Information (RFI)	The General Contractor gets the RFI's from the Subcontractors, evaluate and process them. The CM has to deal with one source of asking questions.	Each Prime send their own set of RFI's. The CM has to manage multiple sources of questions and coordinate the information with the rest of the primes
Damage Claims Back Charges	If any of the Subcontractors cause any damage to the site or to another Sub's work, the GC takes care of that.	In this format, the CM has to step in to judge who is responsible and deal with who pays who for the damage. It gets complicated
Change Orders	One set of change order proposals submitted by the General Contractor who receives and submits the Subs changes order proposals under its name	Each Prime submit their own change order proposals and in case a change includes multiple trades, the CM has to gather proposals from multiple parties

CM at Risk

In this format, the owner asks the construction manager to deliver a project guaranteeing the outcome in time and cost, or there will be a penalty to pay. For example, you have to get the job completed in fifteen months and not exceed $25 million. Obviously, in this case, the construction manager has a much larger risk than the other methods. That means the following:

- The CM has to have more control over the designer and the contractor since the CM is liable for the final outcome.
- The fees for the CM services will be much higher.
- The owner's relationship with the CM is different. After all, if the CM is to be held liable for the outcome, they have to be given more freedom to operate and make decisions.

Guaranteed Maximum

This is the process where the owner enters into a contract with the general contractor where the GC has to complete the project and not exceed a set cost. The general contractor in this case has to study the plans and site very well and account for all potential cost and incidental work.

In this case, there will be a period of studying plans; discussing missed items and some qualifiers have to be identified in the final negotiated price. This method makes the topic of change orders a very delicate discussion. In this chapter, I gave you an idea about each of these methods. There are whole books on each of these topics that should be consulted for more details.

Chapter 8
Bid/Award Phase

At this point in time, you, the construction manager, completed the pre-construction reviews, the architect completed the design, and the design documents are assembled and ready to be advertised.

Before I move on, notice what I called the plans and specifications: Design Documents. The same documents take different names throughout the phases of the project:

• During the design phase, they are called Design Documents.
• When the plans, specifications, and bid instructions are assembled, they are called Bid Documents.
• When you start construction, the whole package is referred to as Construction Documents.

At this point, I would like to emphasize the importance of performing a thorough pre-construction effort to make sure that all details and information needed for construction have been addressed in the design documents. These are the topics that we addressed in the previous chapters.

What Is the Bid/Award Phase?

In short, this is the phase where the owners, with the help of their consultants, the architect, and the construction manager, assemble the project's documents, put together the request for bids, advertise to contractors to submit bids, and finally, award the project to the successful bidder.

Simple? Hardly. Let's get into the details of how it works.

Bidding Private-Sector Projects

Private sector projects, by their definition, are projects funded with private money, unlike public works, which are funded by public funds. Because of that factor, private owners have much more flexibility in how they advertise the project and to whom they award the project. After all, it is their money.

The private owner can be the owner of a single-family house, a private investor who owns multiple shopping centers, or a huge organization with a multibillion-dollar portfolio.

Here is the outline of the process and what is involved:

- Assemble the bid package. This package will include:
  - The plans and specifications.
  - Additionally developed sketches or modifications to the original set of documents.
  - Instructions to bidders outlining the process and forms required to be submitted with the bid.

- Advertise - The private owner may elect to advertise his/her project in a general publication inviting interested bidders to submit bids or may elect to invite few contractors that they know. The private owner may elect not to advertise and simply ask the contractor that they know to negotiate a price for the project.

- Pre-bid meeting - This is more common in public works projects, which I will address in more details below. However, a private owner may elect to conduct this meeting. In this meeting, the interested bidders meet to learn about the project and get specific information they need to know in preparing their bids.

- Request for information - As the bidders are studying the bid documents to prepare their bid, they may have questions to clarify certain information. The process should be defined in the advertising documents giving contact information for such inquiry.

The bidders send their questions to the owner who consults with their design team to clarify the needed information.

To keep the process fair, the owner should publish the questions and answers so all bidders get the same information to compete equally. This is more strictly applied in public works, as you will read below.

- Subcontractors - The private sector general contractors have more flexibility in selecting the subcontractors to use for the project than public works projects. They may also have more flexibility in shopping for better prices and changing subcontractors after the bids have been submitted.

As you will see below, this is not the case in public works projects. The main challenge in getting bids from subcontractors is making sure that they have covered their particular complete scope. The bidder should read the subcontractor's bids very carefully, especially the exclusions. For example, an electrical subcontractor may have excluded hauling the dirt that came out of their trenches (spoils) off the site. That could be a substantial cost that the general contractor may end up paying out of pocket since their bid to the owner did not exclude that part. Or a structural steel subcontractor may have excluded taking the delivered steel off the delivery truck once they get to site.

• Bid Day - The process of submitting the bids, the location, and time should be specified in the advertisement of the project. Again, since this is a private-sector project, the owner has a lot more flexibility as to how to conduct this process. For example, the owner may elect to accept bids any time they want. In fact, if they did not like any of the bids, they may elect to wait for one last contractor to take their time and submit.
• Bid evaluation - The owner reviews the bids submitted to evaluate and select the successful bidder. This is where you, the construction manager, will be asked to evaluate and recommend. This process can involve the following:
 ○ Verifying that the scope is all covered.
 ○ Verifying that the submitted costs are acceptable.
 ○ Possibly, interviewing any number of the bidders to qualify the firm and/or negotiate their prices.

The private owner is free to select the firm that they see best fit for the project. That bid may not be the lowest bid amount.

• Award - Now that the owner selected a contractor, the owner may issue a Letter of Intent (LOI). That tells the contractor that the owner selected your form and intends to award the project to your firm. The letter may ask that the contractor submit some paperwork like the following:
 ○ A copy of their license.
 ○ List of their subcontractors and vendors.
 ○ Insurance certificates
 ○ Bonds (we'll explain this in more details below for public work projects)

Later, the owner issues the official award letter, or simply, both parties sign the Final Contract Agreement.

How General Contractors Bid Projects?

Owners have to respect the fact that there is a cost associated with bidding a project that contractors have to spend on, whether they get the job or not. We discussed in chapter 7 several project delivery methods: Design-Bid-Build, Design-Build, CM-Multi-Prime, etc.

Each of these are bid differently, and in some cases, the contractors have to spend money on submitting a pre-qualification package, hoping they would get on a short list of approved bidders.

If you happen to be an owner, I hope the message to you is to make the process as clear and fair as possible so bidders don't pay out of pocket in an unreasonable manner.

Typically, the process involves the following:
- A thorough review of the bid documents.
- Visiting the site to check access, site logistics, and other site conditions that may impact the cost.
- Consulting the subcontractors for pricing and technical information.
- Figuring out how much overhead costs to allocate for the bid. The hard cost is simply the cost of the work. The soft cost consists of the following:
   - The staff to manage the project.
   - The site overhead, including the temporary site office, site toilets, phones, water, etc.
   - A portion of the main office's operation to allocate to this particular bid.
- Communicate with the owner in case a clarification is needed.
- Putting the price together and submitting the bid package.

That is why it is important for the estimator to be someone who understands how things are built versus simply calculating quantities multiplied by set rates. For example, on a small tight site, there may be a need to remove fences and make temporary wall openings to install an air-conditioning unit, which will cost more than what the estimating books list for such an installation. Or the workers have to park far from the site, and there will be a cost of hours spent back and forth between the site and their tools in their trucks, etc. Bidding is a risky task, and the contractor better think of all costs, or they will get stuck with the submitted numbers and end up at a loss. In the private sector, there is usually more flexibility with time to prepare the bid, communicating with the owner, and negotiating the final bid amount.

Bidding Public Works Projects

This is a whole different game than the private sector. Public bids are governed by the Public Contract Code, which may be different in a another state. We advise you to familiarize yourself with those regulations if you are a professional dealing with a public works project. My business is conducted in Los Angeles, California. So the outline below is taken from this perspective, but will generally explain to you what is involved.

Let's take the above tasks outlined for the private sector and see how differently it is done if the project is a public works project.

Again, each delivery method may be bid differently per the instructions advertised with the invitation to bid. In the following section, I will outline the process involved in the most commonly-used method, which is Design-Bid-Build .

Here is the outline of the process and what is involved:

• Assemble the bid package. That is basically the same as the private sector. This package will include the following:
  ○ The plans and specifications.
  ○ Additionally developed sketches or modifications to the original set of documents.
  ○ Instructions to bidders outlining the process and forms required to be submitted with the bid.

• Advertise. Advertising a public works project is more regulated and has more restrictions than private-sector projects .
  ° A public notice of at least two consecutive weeks or once a week for more than two consecutive weeks if the longer period of advertising is deemed necessary by the department, as follows:
  (a) In a newspaper of general circulation published in the county in which the project is located.
  (b) In a trade paper of general circulation devoted primarily to the dissemination of contract and building news among contracting

and building materials supply firms.

(c) The department may publish the notice to bidders for a project in additional trade papers or newspapers of general circulation that it deems advisable.

○ The advertisement will identity the date and time to submit the bids and, if applicable, the date of a pre-bid meeting.

○ Public bids are typically submitted in sealed envelopes that are opened in public where the results are read to the present bidders and later published for the public to read. The bids shall be accompanied by one of the following forms of bidder's security:

(a) Cash

(b) A cashier's check made payable to the school district.

(c) A certified check made payable to the school district.

(d) A bidder's bond executed by an admitted surety insurer, made payable to the school district. Upon an award to the lowest bidder, the security of an unsuccessful bidder shall be returned in a reasonable period of time, but in no event shall that security be held by the school district beyond sixty days from the time the award is made.

• Listing of subcontractors. This is a very sensitive issue and heavily regulated by the Public Contract Code . The bid documents usually have a form for the general contractor to list all their subcontractors that are performing work whose value is equal or more than half of one percent of the total bid amount. If a subcontractor was not listed for a certain trade, then the general contractor is saying that they will perform that trade with their own labor. Of course, they have to be properly licensed to do so.

These regulations are also meant to avoid bid shopping or collusion. The public is given here a fair and equal opportunity to provide their best bid. If the general contractor used the bid of a concrete subcontractor, for example, to be the lowest bidder and win the project, they can't go shopping for a lower number afterward.

Collusion means that the general contractor can't list a sham name of a subcontractor when the real performer of the work is another firm.

So does that mean that the general contractors are stuck with the listed subcontractor? What happens if the listed subcontractor can't perform the work? How about if they turn out not to be properly licensed? What if they decide to change their bid amount to the general contractor after the bid opening? What if they can't provide the required bonds, etc.? On the other hand, what if the general contractor got a lower concrete bid and wants to use the other subcontractor that was not listed?

° Substituting listed subcontractors. To avoid any unethical behavior of abuse of the process, the public contract code regulated the process of substituting subcontractors. In summary, here is how it goes:

(a)   The general contractor submits a request to the awarding agency to substitute a subcontractor listing the reasons why. The public contract code has a list of legitimate reasons.

(b)   The awarding agency sends a notification to the relevant subcontractor that the general contractor has requested their substitution.

(c)   The subcontractor has five days to respond. If they agree to be substituted, the issue is resolved, and the new subcontractor will be added to the project.

(d)   If the subcontractor protested the request, a hearing is set where the awarding agency listens to the general contractor and the subcontractor to learn about the arguments of both parties.

(e)   As a result, the awarding agency may accept or reject the request.

This process can turn into a legal fight with lawyers involved. So don't take this lightly.

My business is in California. If you operate in another area, please check your local requirements. An appendix in the end of the book titled *Review of the Public Contract Code* includes some excerpts from California and explains the objectives of these regulations. I advise you to go through it to get an idea about relevant regulations.

• Pre-Bid Meeting.

A pre-bid meeting is a meeting with all the potential bidders and the owner's team to discuss the specifics of the advertised project before submitting the bids. Here is how it works.

Typically, the owner announces the date and location of the

pre-bid meeting in the advertisement. Most owners call for a mandatory pre-bid meeting, and some call for a non-mandatory meeting.

If the meeting is mandatory, then only the firms that attended the meeting can bid on the project. If the meeting is non-mandatory, then any firm can submit a bid.

Who Attends This Meeting?

General contractors, subcontractors, vendors, the owner's team, the architect/engineers, safety officers, labor compliance officers, the inspector, the end user of the new facility, and other people with stakes in the project.

How Is This Meeting Managed?

If you are a bidder, you better be there on time, or you may risk the possibility of not being allowed to attend even if you are one minute late. I advise you to always arrive at least a half hour early. I usually arrive an hour early. You have to account for possible traffic, getting lost, finding parking, or walking a long distance between parking and the meeting place.

If you are on the owner's team, then you should be there very early to make sure the meeting location is ready; you may need to make copies of the agenda and other paperwork.

So the bidders arrive and sign a sign-up sheet, listing their name, their firm's name, and contact information. If the meeting is mandatory and your firm's name is not on that list, your bid will be rejected.

I've seen some owners ask firms to sign in and out at the end to make sure that bidders attended the whole meeting.

The sign-up sheet gets copied and distributed or posted online so all bidders know who else is bidding. Subcontractors use that list to send bids. Owners use that list to double-check that a submitted bid was from an attendee.

What is Covered in This Meeting?

The typical topics are the scope, phasing, staging, administrative requirements, and it is a chance for the bidders to visit the site and become familiar with it.

Questions and discussions come up during the meeting. It is very important to keep in mind that nothing verbal holds ground in the bidding process. The bidders will bid based on what it is printed in

the bid documents. In fact, the moderator of the meeting will emphasize this point. In other words, a contractor can't use a verbal statement from the pre-bid meeting as a basis for a contractual dispute
during the construction phase.

• Pre-Construction Requests for Information

Questions will come up during the pre-bid meeting and during
the bidding phase. In public works, it is very important to keep
in mind that all bidders should be equally informed. That is why
questions and answers have to be in writing and made available to
all bidders.

The vehicle to do that is to issue an addendum.

An addendum is a document that lists all changes from the original bid documents, including the questions and answers from the
bidders. The addendum is posted online for all bidders to see. If the
pre-bid meeting was mandatory, then an email goes out to all firms
on the list, making them aware that an addendum was issued. If the
bid was not mandatory, then the owner usually posts the addendum
online for all the public to see.

The bid documents will have a space for bidders to
acknowledge receipt of all addenda. In public works, if a bidder did
not acknowledge all addenda, the bidder will be considered nonres-
ponsive, and their bid will be rejected.

If you are an owner's representative, and you get a call from a
bidder, make sure that you don't provide an answer that will place
that particular bidder at an advantage over others. That may become
a reason for other bidders to protest and cancel the bid. If that happens, the owner has to go through the whole advertisement process
all over again.

• Bid Day

The bid day for a general contractor bidding a public works
project is a very tense time. Granted, the general contractor took
time to study the bid documents, run their own estimates, and talk
to their subcontractors, but when it is time to put down the final
number and have to be the lowest "responsible" and "responsive"
bidder, the word *tense* is too soft for that last fifteen minutes before
bid time.

Here is what is happening:

- The bid time is 2:00 p.m. for example.
- The contractor filled all the bid forms a day before or on that same day. Someone has to verify that not a single form or signature is missing; otherwise, it is grounds to disqualify the bid.
- The contractor sends a representative to the bid opening room with the forms and the envelope that should be submitted sealed before 2:00 p.m. You submit the envelope at 2:01 p.m., your bid won't be accepted.
- Meanwhile, the team at the office is working their estimate sheet while getting tons of faxes from various subcontractors. Some they know and some they never heard of before.
- Few hours before the bid time, the team has time to examine what was included and what was excluded in the faxed bids from subs.
- Now, we get to the last fifteen minutes; suddenly a bid comes in for plumbing, for example, that is really lower than the selected plumbing bid. What do we do? Shall we list this new sub? Can someone call this guy and find out if his scope is complete? We don't have much time; we need to give our guy at the bid room the final list of subs and the bid amount. Someone may yell, "Go ahead and list this new plumbing sub!"
- Now we are seven minutes before 2:00 p.m., and we don't have bids for the site concrete and the elevator. Someone has to make a call about how much money to put for these items. In some cases, these got estimated, and in some cases, the contractor just takes a risk and plugs a number based on their best guess.
- "Hey guys, there is no more time; we need to give our guy the final list of subs and bottom line number."
- So the team managed to do that. The guy at the bid room manages to write down the information, seal the envelope, and submit on time.

Now the storm has passed, and the team at the office sits down like they just went through a tornado, and the weather got calm again. They are waiting for a call from their guy at the bid room.

Let's go to the bid room. The public owner opens the submitted envelopes and reads the bid amounts in public. All the bidders' representatives are there writing down the results.

Now the call comes in from the contractor's representative with the results. Have you ever put a coin in a slot machine and sat there watching the pictures roll, not knowing what the result is? It is the same feeling.

Hey guys, we got the job, or we came in second by $5,000, or we were seven out of fourteen bidders, or any other results. What happens next?

If the contractor was the lowest bidder, that firm will check their bid numbers to make sure that nothing was missed. Especially if their bid was $14 million and the second bidder was $15 million. Then they need to make a decision to take the job or withdraw. Mind you, there are a limited number of days to decide to withdraw.

If the contractor was not the successful bidder, they may say, "Oh well" and go to the next bid, or the contractor may decide to examine the bid documents of the lowest bidder to check if that firm missed some information or did not comply with some requirement, especially if the bidder was the second. They may be able to find something wrong with the first bidder and protest to get them off the job.

I've seen firms that placed fifth and looked for reasons to protest and forced the owner to cancel the bids, thinking this will give them another opportunity to bid on the project.

No wonder contractors call the room where they prepare bids the War Room .

Best Practice

So many public works contractors lose their business or get into legal trouble if the bid wasn't properly prepared. So here is an outline of sound steps to take while preparing your bid:

- Give yourself enough time to study the plans well.
- Run your estimate so you know how to evaluate the subcontractors' bids.
- Talk to subcontractors early about the project.
- Prepare the bid documents a day before.
- Visit the site to incorporate site conditions.
- On bid day, examine the incoming bids, especially their exclusions.

Chapter 9
Construction/Closeout Phase

Now that the previous pre-construction tasks are completed, the project enters the world of reality; that is the construction phase. This is the phase where all that was missed during design will surface, and all the good effort done will pay back.

You can be a construction manager handling projects, an owner's representative, or managing the project for a contractor. Either way, you need to have a deep understanding of the concepts presented before and in this chapter.

I will try to cover all the issues that will be encountered during this critical phase.

Typical Staff on a Construction Project

As a construction manager, you will deal with a wide variety of people. It is important to understand each person's position and the role of that position.

Typical Owner's Team:

The owners vary in size and organization. You may be dealing with a single person as an owner of a house up to the federal government as the owner of the project. The owner can be a small city or a large school district.

Whether you are working as a construction manager/owner's representative or for a contractor building the project for the owner, you need to understand the organization structure of the owner. For the purpose of this chapter, let's take the owner of an average-sized organization . Here's a list of team members:

1. Head of the Owner's Organization: The party that owns the project and most of the time funding the project either privately or publicly.

2. Owner's Construction Manager: This is the authorized manager who will head the team of a particular project or projects. This

person will be the main contact point for the contractor.

3. Architect: The main design professional who designs the architectural portion of the work and coordinates the design with other engineers, including the structural, mechanical, electrical, landscape, hardware, and interior design, etc.

4. Inspector: The main inspector or the "Inspector of Record" is in charge of making sure that the construction is per the approved plans and specs. In some cases, public entities have their own team of in-house inspectors. In other cases, like the private sector, the city or the government agency with jurisdiction performs the inspection or final signoffs, and it can be both.

5. Labor Compliance Officer: In public works, the contractor and subcontractors have to pay workers set prevailing wages as a minimum. These rates differ with the trades. The labor compliance officer makes sure that this is taking place. He/she may go to the site and do random interviews with workers to make sure they are properly paid. If not, the relevant contractor will be assessed a penalty.

6. Safety Officer: This person is in charge of making sure that the site and the contractor are operating following the set safety regulations.

7. Environmental Lead: Some projects may involve abating asbestos and lead. This trade is heavily regulated and requires licensed people to perform the abatement. This person makes sure that all the proper equipment and steps are taken to perform this task properly. I have seen school projects shut down because a teacher saw white powder on the ground and thought it was asbestos. It turned out to be gypsum powder. But because the abatement sub did not clean properly after abatement work a day before, the school shut down in fear of liabilities and risking the students' health.

8. Commissioning Agent: Commissioning is the process of making sure that all installed systems are functioning per the intended design. That is Heating, Ventilation and Air Conditioning (HVAC), electrical, security, etc. This agent is the person in charge of this process on behalf of the owner.

9. LEED Agent: LEED, or Leadership in Energy and Environmental Design, is a green building certification program. Some projects require a LEED certification, a process depending on getting credit point for different levels of LEED certificate. The LEED agent is the person in charge of managing the process for the owner.

10. Community: Some projects have the local community or

the end users involved in the process of building certain projects. It could be for political reasons or simply a local interest. Typically, there will be a committee representing the community involved. They may attend meetings. The owner's representative will handle that communication.

Typical Contractor's Team:

1. Project Manager: The one in charge of overseeing the progress of the project dealing with the subcontractors, the owner, the architect. In short, this person is the focal point of contact for the contractor on this project. His/her tasks include managing the budget, the schedule, and the contractual issues with all other parties.

2. Project Engineer: This person typically prepares a variety of paperwork, including change orders, Request for Information (RFI), and submittals. He/she does the legwork for the project manager to take the appropriate decisions.

3. Safety Officer: This person is in charge of making sure that the site and workers are complying with safety rules and procedures, safety meetings and reports, and will be the person talking to OSHA in case they show up to the site in case of an accident or simply checking safety compliance.

4. Project Accountant: This person logs in all expenses and earnings to keep proper accounting records of the project's financial status. He/she provides periodic reports to the project manager.

5. Superintendent: This is the site supervisor who runs, supervises, and coordinates the work on site. He/she coordinates with the subcontractors and vendors, verifies that the work is per contract documents, handles inspections, and provides feedback to the project manager.

6. Estimator: Performs estimating for change orders or other tasks that need estimating on the project. Evaluates the proposals from the subcontractors. Participates in relevant negotiations.

7. Scheduler: Prepares the baseline schedule and the monthly updates. Produces the needed reports regarding the schedule as required by the contract documents.

8. Administrative Assistants: Handle document control, filing, and organization of the project's files.

Contractor's Subcontractors

First, let's start by defining who is a subcontractor. The owner signs a contract with the general contractor to build the project, which includes all the involved trades: Concrete , plumbing, electrical, painting, etc. This is called a prime contract. Unless the general contractor is doing all the work with their in-house labor, they will have to hire specialty firms to do the concrete, plumbing, and other trades. These firms carry specialty licenses for these trades and will sign a contract with the general contractor, not the owner. That is why these contracts are called subcontracts and the firms are subcontractors.

Should one of these specialty trade firms sign a contract directly with the owner, they become prime contractors themselves.

Most of the general contractors, especially on public works projects, hire subcontractors. The process of selecting the subcontractors is different between private-sector jobs and public works.

In the private-sector projects, there is much more flexibility in selecting the subcontractor that the general contractor wants and in negotiating the cost. In some cases, the private owner requests a specific subcontractor that he/she likes.

In public works projects, it is a whole different case. These projects are subject to the Public Contract Code. So please check your local regulations on this topic. To give you an idea, I'll talk about how this is done in California. The public contract code requires that the general contractors list in their sealed bid forms all the subcontractors that are performing work more than half of one percent of the total bid amount. If the general contractor does not list a plumbing subcontractor, for example, then they are declaring that they are licensed to and will perform this work in-house.

Additionally, if the general contractor wants to change one of the listed subcontractors, there is a set procedure for that, which we touched on in previous chapters.

What I would like to explain in this chapter is how to best manage and deal with subcontractors because they can make or break the project.

Real Life Gauge

This has been a recurring theme of this book, which is that having good relationships makes all the difference in the world. I believe in long-term relationships. There is great value to be gained. If you are a general contractor with a good relationship with your concrete subcontractor, for example, there will be more cooperation on cost, problems on site, etc.

So what does that mean? How do we do that?

Well, there is a money side of this and a bit of personal interaction. First of all, firms don't do construction for their pleasure, like other business; they are in it to make money. Construction is a very risky business, and cash flow is key. Suppliers have to be paid before the owners pay, laborers have to be paid weekly before the owner pays, etc. So the general contractor stepping in to pay some or all of these invoices before they collect from the owner is part of building this relationship.

Another situation is when there is a problem on site or with the owner. If the general contractor stands by the subcontractor instead of immediately passing the blame and threats to them, that also contributes to building that relationship.

So similar to any relationship, it takes an effort to build that trust. General contractors who do not understand this value will pay a heavy price on the job.

Project/Relationship Management

Previously, we listed the typical players involved in a project. As a construction manager, whether for the owner or the contractor, one of the main challenges is the human interaction and dealing with the various personalities and agendas.

I have seen so many projects where, because people didn't get along, the plans and specs turned into weapons used to get back at each other. All fights safely covered under the title "This is Contractual" when they are actually egos, personality clashes, defending liabilities, or simply insecurities trying to self-edify.

On the other hand, I have had many projects where, because of the good relationships, everybody went out of their way to help resolve matters, and the plans/specs became documents used to justify the solutions.

No matter what your position is in the project, try your best to establish good relationships with others. You can do that by understanding the fears and touchy issues of the others involved and help them feel safe, more comfortable, and appreciated. Hopefully, others will treat you the same way.

Okay! Enough philosophy; just what are some of the typical problems that put the team members on a collision course?

Owners

Owners have their own communities that have stakes in the project. It could be the funding source, public constituents, end users, board members, etc. It can also be deadlines and budget limitations.

Based on the above, owners may do the following typical mistakes that end up making the project cost more and take more time, which is in direct conflict with the very same liabilities they are trying to prevent.

• Rush to advertise the project for bids before the plans and specs are properly prepared or reviewed for errors and omissions.

• Assign a limited budget to the architect, lesser than the real cost of design, which results in many change orders and claims and a frustrated architect who wants to do a good job but now is spending out of pocket due to the small allocated budget.

• Hire a construction manager and transfer all their pressures and blame to the CM, who finds himself/herself in a defense position against his/her client (the owner).

• Advertise the project with a very tight duration, which places the contractor scraping for time extensions, claiming delays on every issue, which in turn places the whole team in an adversarial position.

• Not performing a thorough pre-construction review, which will detect the potential claim and change order issues for budget purposes. To the owners' peril, a lot of them don't realize that the cost of this service is much less than the cost of the potential claims and disputes.

Architects

Architects have to manage the designs by their people and the other engineers like structural, civil, electrical, etc., while keeping the hours within the allocated budgets. Because of the owner's imposed low budget and short design period, the designs get rushed, resulting in poor coordination among the various trades. The result is change orders, claims, and animosity.

The project gets delayed, which means more months on the job for the architect to support the construction phase. If the owner doesn't want to pay for this extended period, you can imagine the architect's position, who is now working for free.

Contractors

There are the firms that get the plans, the final result of the above messed-up scenario, and they are expected to give a final lump-sum price for the project.

When errors and omission questions come up, the owner feels the pressure of having to pay more, and the architect feels the pressure of explaining to the owner why these details were missed. Of course, the architect won't tell the owner that it is because of his poor planning and allocating an unreasonably tight budget. The result is clash, claims, and animosity.

The subcontractors will carry a lot of this burden since most jobs are subcontracted; their numbers were possibly squeezed down to a minimum.

Construction Manager

This is a consulting position where the expectation is that they will perform magic and make all the above work just fine, all because they were hired as a construction management consultant. The CM has to navigate the project through the above maze to get this project to completion hopefully to meet its time and budget.

Okay, so now I managed to paint a gloomy picture, and you are probably asking yourself, "What does all of this have to do with the title of this chapter: 'Project/Relationship Management'?"

Reading the above just described to you the pressures that each party is under because of their unique position in the project. These pressures drive each of these players to behave and make decisions to protect their own issues. Decisions that may sound unreasonable to one can be perfect for the other person.

The tone is set by the owner. The owner has to understand the pressures he caused everyone due to the time and budget constraints that he imposed and be ready to accept the fact that this project will have change orders and potential time extension. The owner has to make the others feel that if these claims are fair, they have nothing to fear from it. A good relationship with the architect and the construction manager from the start will take care of that.

Once that tone is set and the construction manager feels that they are not going to be blamed for everybody else's mistakes, they will be able to assess the change-order requests in a fair manner.

That places the contractors in a position to focus on getting the project completed versus legal self-protection at every corner. A good relationship between the owner and CM and a good relationship between the CM and the contractor will take care of that.

Of course, the more people are involved in establishing good relationship helps the project and the parties involved. I understand this is not a magical solution, but working in that direction is worth it.

So if you are working as a construction manager for the owner or the contractor, it will be wise to understand the various roles of the above people to think about the best way to interact and cooperate to help make the project a success.

This is not always easy. The human factor is the toughest challenge in managing projects.

Job Start Meetings: Notice to Proceed (NTP)

So now the contractor was awarded the project, they filed the insurances and bonds, got somewhat ready by signing up immediate trades' subcontractors for the immediate trades.

The owner will typically ask to have a pre-construction meeting. In this meeting, all parties that have any stake or say in the project will be present. After this, only few will be around for the duration of the project on a regular basis.

Who Attends This Meeting?
  • General contractor's project team. The project manager, project engineer, superintendent, and upper management from the main office.
  • Sometimes subcontractors attend.
  • The owner's construction manager.
  • The rest of the owner's team listed in the previous section.

What Is the Purpose of This Meeting?

In this meeting, the owner's representative takes the lead and goes over a set agenda, including all issues that the team members need to be aware of or discuss before the start of the project. Here is a list of typical discussion points:

1. Introduction and Background

a. Attendees sign meeting attendance sheet. This is an important document to prove that someone was there and heard the presentations and needed information. The owner's representative distributes this to all attendees for their record.

b. Introductions

i. Owner representatives (the construction manager, inspector, architect, etc.) introduce themselves.

ii. The contractor's team introduces themselves.

c. Roles and responsibilities: To me, this one of the most important topics. The construction manager defines the role of each of the attendees and the procedures and channels of communication. If this is not defined from the start, there will be chaos.

2. Safety

a. Review roles and responsibilities for site safety; introduce contractor safety representative.

b. Review project safety standards, Cal-OSHA construction safety standards, and contractor responsibilities.

c. Review contractor safety program (first aid, fire protection, sanitation, drug testing).

d. Submittal/receipt of contractor site-specific safety action plan.

e. Review worker orientation and training program.

f. Review workplace conduct.

i. Alcoholic beverages are not permitted on the project site.

ii. Contractor to ensure employees are dressed properly and to not use profane language.

iii. No smoking is allowed on owner property.

g. Restrict operation areas and traffic control.

i. Contractor shall provide proper flagmen when impacting street traffic and crossing sidewalks.

ii. Construction debris shall not accumulate and shall be removed as required by contract documents.

iii.Barricades to be installed as required to ensure the safety of pedestrians.

iv. Contractor to control and mitigate dust caused by construction activity as required by contract documents.

3. Environmental

a. Review contractor's proposed usage of hazardous materials.

b. General hazardous material handling.

c. Asbestos and lead reports, coordination with environmental consultant.

d. CEQA mitigation measures and reporting requirements, if applicable

e. SWPPP permitting, reporting, and completion notice requirements

f. AQMD permitting requirements (generators, boilers, etc.).

4. Construction

a. Review contract scope of work.

b. Temporary facilities

    i. Lay down area and storage areas.

    ii. Temporary offices/trailers for the construction manager and inspector.

    iii. Power, water, gas, sewer, storm drain, Internet

    iv. Portable toilets

    v. Equipment (computers, copiers, etc.)

    vi. Trash removal

    vii. Site access (including deliveries)

    viii. Parking availability (on-site and off-site, restrictions, etc.)

    ix. Housekeeping

    x. Security

    xi. Contractor's normal working days.

    xii. Contractor's normal working hours.

c. Coordinating work areas with others

d. Permits requirements

e. Environmental impact statement (if necessary)

f. Work coordination

g. Quality control and inspection

h. Work by separate work contract

5. Labor Compliance

a. Discuss labor compliance program requirements, in particular the following:

    i. Submittal of required labor compliance documents.

    ii. Prevailing wages, violations, certified payrolls, and associated topics.

b. Review contractor responsibilities and remedies:

    i. Identification of "problem" subcontractors.

    ii. Reviewing and verifying subcontractor electronic submittals for compliance.

      iii. Review of hearing and appeals process.

6. Construction Schedule and Progress Reporting

    a. Review required schedule submittals.

      i.   Baseline schedule requirements.

      ii. Schedule updates requirements.

      iii. Important phasing and milestones issues.

    b. Critical work sequencing

    c. Progress meeting, look-ahead schedules.

7. Submittals

    a. Status/schedule of Survey of Existing Conditions form submittal. This is where the contractor and the owner inspect existing damages before start of construction. This is done so the contractor does not get blamed for any damage they did not cause.

    b. Submittals

      i.   Schedule of submittals.

      ii. Submittal procedures and protocol.

      iii. QA/QC procedures

      iv. Safety action plan

    c. As-Built Drawings

      i.   Contractor must keep "as-built" drawings current with installation of the work.

      ii. Owner will review "as-built" drawings prior to each monthly pay app approval and submittal for processing.

    d. Shop Drawings, Product Data, Samples

      i.   Shop drawing submittals, reviews, and schedules

      ii. Drawing control (revisions and distribution)

    e. Closeout Documents

8. Testing and Inspection

    a. Contractor to provide information and access for testing and inspection as required.

    b. Calling for inspection process.

    c. Materials test report distribution.

    d. Deficiency Notice – avoidance and clearance

    e. Punch List

9. Quality Assurance/Quality Control (QA/QC)

    a. Contractor responsibilities

    b. Inspector's role

    c. Prefabrication submittal requirements

10. Contractor Payments

a. Review of applications for payment process.
    i.   Progress payment schedule and progress measurement.
    ii.   Schedule of values.
    iii. Submittals required prior to payment.
    iv. Owner assessment process.
    v.   Allowance disbursements process.
b. Final pay application
    i.   Certificate of Substantial Completion process
    ii. Conditional/unconditional release and waiver upon final payment.
c. Release of retention
    i.   Timing
    ii. Retention to withholds
    iii. Punch list withholds
    iv. Other withholds, claims, and assessments.

11. Contract Management

a. Correspondence
    i. Contractor shall send all correspondence to the construction manager's/owner's representative. It is very important to have one focal point of communication to avoid confusion and have better coordination.

b. Contract Compliance
    i. Contractor/subcontractors to review contract documents prior to starting work.

c. Dispute Resolution Process

d. Claims Process

12. Questions or Comments

Real-Life Gauge

When I worked as a construction manager/owner's representative, I felt that this meeting was the most important meeting of the project, for many reasons:

1. Think about this: Here we are in a room with people who may have never met before, and they are about to engage in a team effort for the next sixteen months to build a project. Each one is concerned about their part of the liability, agenda, people to report to, etc. People don't have any idea about the other's background, personality, etc. This meeting, to me, was my opportunity to break the

ice and inspire an atmosphere of positive teaming to give a comfort level to all players that this will be a positive endeavor.

2. There will be people in the meeting whom we may never see again, and some will be there on a daily basis. This is the perfect time to set the rules, procedures, define everybody's role and responsibility. For example, I emphasized that I will be the point of contact, and all correspondence to and from the owner will go through me. Having everybody there hearing this will prevent redundant communications or giving conflicting instructions. I would stress to the contractor that no one but me authorizes change order work. Again, having everybody there prevents others from instructing the contractor to do added work and informs the contractor that no change order will be paid unless the additional work was authorized by me.

3. A team is like a car engine. Each part does its role and only its role. The parts are connected so the car runs smoothly, which will not happen if any of the parts is dysfunctional. Well, the car's team structure is much easier to manage since these parts don't have human nature problems, and they will do what they were designed to do. With people, the challenge is bigger due to human nature. So to cut the philosophy, here are the best ways for a team to work together:

a. Each team player understands clearly their part.

b. Stick to their role while helping the other member succeed without infringing on their role.

c. Have one captain running the ship, who can delegate and coordinate.

That is why I felt the preconstruction meeting is the most important meeting of the project. It is a great opportunity to break the ice, meet the new people, and inspire a positive feeling before we start the work.

The same is true if you were the construction manager for the contractor. This is the time that you give the owner a great first impression and start a relationship with the owner to have a smoothly running project.

I have been to some of these meetings as a project manager for a contractor, where I left the meeting knowing this project was going to be a nightmare. The owner's representative was mostly threatening, yelling, emphasizing on what is not allowed, etc.

Remember, teaming and people management are the key to successful projects.

Partnering

I am a big fan of this effort. Partnering is an event held by the owner or the contractor, before the start of the project, with all the people involved in the project to break the ice, get all parties to know each other informally, and create a positive atmosphere.

No, it is different from the preconstruction meeting. That meeting is mostly technical. This meeting may include technical topics, down to a silly game that builds feeling of trust between the contractor's project manager and the owner's representative, for example.

There are professional businesses that provide this service; that is, setting up and administering the event, the agenda, the topics, etc. Of course, if you add food, it helps. The meeting can start with people introducing themselves, their backgrounds, maybe sharing some personal information about their hobbies, how many kids they have, etc. Suddenly the contractor, the inspector, the architect are human beings! Hey! They have kids and hobbies like you. Surprise, that subcontractor shares your love for hiking.

The agenda may include discussing typical sensitive problems that occur on projects and brainstorm on the best ways to handle them. Parties may offer commitments for what they will do to help resolve conflicts.

This event may be a half-day or even full-day event. I preferred the full day because it meant lunch was on the agenda, and if the event was in a hotel, well, even better.

Now, people know each other personally. Instead of the typical statements during construction like "The architect rejected this" or "The inspector has to be called first on that," now it is "Let me call John [the architect], and he'll work with us on this." Get the message?

So is this the magical solution? It all depends on the attitudes and personalities of the parties. If there is intent to work things out, then the event will produce the needed results. On the other hand, I've seen people who felt this was another waste of time and that they got dragged to this nonsense, and couldn't wait until it was over. Those people go in and out of this event without any change in their mindset. They will go about doing their work the same way.

In conclusion, attitude, attitude, attitude is the driver. If you see the benefit of having a smooth project, then do the effort to work on teaming. Be fair, be nice, be honest, be straightforward, and don't play the typical games of finding loopholes to stick it to the other side. Help others to succeed while being very aware of your rights.

Submittals

Submittals are the required set of documents that the contractor submits to the owner/architect to approve prior to start of that particular trade.

Typically, the specifications for each trade or scope of work list exactly what is required for that trade. The contractors have to be familiar with these requirements to submit the required items. For example, the contractor won't be allowed to start the ceramic tile work until the submittals have been approved.

Here are some typical examples of required submittals:

Concrete Masonry Unit
Product Data
Samples
Mock up
Material Certificates
Mix Design
Statement of Comprehensive Strength

Structural Steel Framing
Product Data
Shop Drawings
Qualification of Installer
Welding certificates
Paint Compatibility Certificates
Mill Test Certificates
Product Test Reports
Source quality-control reports

Ceramic Tiling
Product Data
Shop Drawing
Samples – Color options
Qualification Data
Master Grade Certification
Product Certificates
Material Test Reports

The contractor collects the needed information for each trade and packages each trade in a separate submittal package for the architect's review and approval before the work of that trade can start.

If the contractor does work without an approved submittal, then the contractor is taking the risk of being rejected and replaced at their cost.

The timing of submittals and the speed of the owner's review can have drastic impacts on the schedule, which can be the subject.

This is a task that should be completed as early as possible in the project. If you are a contractor, make sure you read the submittals specification section to be familiar with what is required. Pass it on to the subcontractors so they don't send you incomplete packages that the architect will reject.

Substitutions

A substitution is when an alternate product is submitted instead of what is specified. Let's go over some typical reasons for substitutions:

• The specified material or product no longer exists. In this case, the architect has to collaborate with the contractor and keep in mind the potential cost difference between the original material and the suggested one.

• The contractor's subcontractor is a representative of a certain model and can provide the suggested alternate at a lower cost to the owner. In this case, the owner can elect to benefit from the saving or simply reject the request for substitution.

• The contractor bids the project with an alternate product, assuming that it will be approved. In this case, the owner has no obligation to accept the alternate product just because the contractor assumed something without the consent of the owner.

In the private sector, the owner and architect are in full control to accept or reject a substitution request. In public works, substitutions get a bit testy. Being a public project, which is supposed to provide an equal opportunity for all qualified parties, public owners can't provide one manufacturer the only opportunity to provide their product if there are other equal products that can perform the same.

For example, a school district may want only Carrier HVAC units because they are used to maintaining those or for any other reason. Other brands like Trane, for example, may object, claiming that their product also complies with the specifications.

So a typical public works project specification will state that when a manufacturer is specified, it is implied that "an equal" product can be considered.

Some public owners write "no substitution is allowed." If you read that, please consult with your lawyer if this will stand.

Some owners write a specification that fits exactly one manufacturer in mind and makes it difficult for another to claim that their product is equal. I have seen claims filed on this issue. Again, please consult with your lawyer on this matter.

Some owners specify that no substitution will be considered after fifteen days from awarding the project. They know that it will take more than fifteen days for a general contractor to sign up all the subcontractors, and by that time, they will have a legitimate reason to reject the "equal" submittal. Please consult with your lawyer on this issue as well.

So if an owner rejects a submittal for an equal product, it has to be backed up by a technical reason that is in conflict with the project's specifications. The owner can't simply say, "Your product is equal, but I want the other one." This may also generate claims.

The Review Process

Each owner has a different method of processing submittals. Some will take them electronically, some request eight sets for each submittal in print, and some may have a project management software to use for the submittals.

The main challenge to the general contractors is the timeliness of the subcontractors to submit complete packages fast.

The frustration of the architects is receiving incomplete packages. The reason is that they have to waste time on reviewing them to point out what is missing, and then reviewing them again when submitting complete packages.

On the other hand, architects, please don't let your frustration guide your actions. I have seen critical submittals of long lead-time items returned as rejected because of one missing useless piece of information, under the argument of "the submittal was incomplete". You know, you can always return a package with comments and stamp it "approved as noted, no need for resubmittal".

In conclusion, architects and contractors, just help each other out for the sake of the project. If you happen to be a construction manager for the owner, you can play a big role in facilitating a common-sense approach.

Real-Life Gauge

If you talk to the lean construction advocates, their ideal world would be a project without any submittals. Well, there is some common sense to their logic. For example, if the architect specified TypeX gypsum boards, why do you need a bunch of papers from the contractor about TypeX gypsum boards? Why can't the contractor proceed without this submittal if they are going to use the specified item?

The counterargument from the group that loves submittals is that it is a way for the contractors to tell the owner what they are planning to use on the project. Additionally, what if the contractor is planning a different type of gypsum board?

So what do you think? Should submittals be eliminated, submit everything, or somewhere in between?

My take is that there should be a practical approach that serves the purpose of submittals without unnecessary wasted efforts and useless papers. Processing and logging papers needs time and resources, which is a cost issue.

The thing is that there are many scenarios relevant to submittals and the need for them. Here are some examples:

• I think if the item is a straightforward item, like using the specified Type-X gypsum board, where the contractor will provide exactly what is specified, there is no need for a submittal.

• If an item is indicated to be a deferred approval or to be design/build by the contractor, obviously, there is a definite need for a submittal to be approved by the architect prior to starting the work. Typical examples are fire sprinklers, fire alarms, and structural steel, or it could be a specified system without much installation details shown on the plans, and the architect wants to see the contractor's intended method.

• If the contractor plans to deviate from the contract documents, for example, wanting to submit a different model of an HVAC unit, then obviously a submittal is needed for approval prior to purchasing the unit or doing work on site.

The other sensitive issue about submittals is timing and schedule impact. When a contractor is awarded a new job, it usually takes a while before all the subcontractors are on board because there is a lot of paperwork to be submitted by the subs before signing the contract with the general contractor. There may be contract clauses that are being negotiated, or maybe the price has not been finalized yet.

I have seen specifications where the owner requires all submittals to be submitted within thirty-five days of award of contract. First, this is a very unrealistic request and shows that whoever wrote this has no clue how real life works in the world of general contractors and their subcontractors. It also places the general contractor in a potential schedule default before anything starts. How about that for a positive way to start a project?

Some owners defend doing that by saying that they are trying to avoid any material substitutions by rushing the contractors to submit. I think specifications should be more compatible with real life and the reality of how things actually take place.

My philosophy is making sure that everybody succeeds. For example, you can specify in the contract something like this: "Contractor is responsible to submit all submittals in a timely manner to comply with the approved baseline schedule." Or specify a reasonable time for the submittals, or for the whole duration for that matter. I have met several owners who specify nine months for a project that should take twelve months. Well, get ready for tension and a delay claim every step of the way.

Best Practices

If you happen to be general contractor:

• Expedite signing up the first trades on the project and get their submittals to the owner as soon as possible.
• Your next priority should be the long lead-time items like doors, frames, hardware, structural steel, cabinets, etc.
• In short, getting all needed submittals in.
• When you first get the job, go over the specifications and list down all needed submittals in a log.
• Have a log that tracks the time you received submittals from the subcontractors, sent to the architect, and back. Have a column

for the status.

• Keep this log visible to you on a daily basis until all submittals have been approved.

• The wheel that squeaks gets the oil. Be a squeaky wheel and bug the subs and the architect until all submittals are done. Remember, your project may be your priority but not everybody else's.

If you happen to be owner's representative:

• Set up a practical procedure to process submittals between the contractor and the architect.

• Specify realistic durations in the contract for submittals review. A three-month job isn't going to get done if you give the architect twenty days to review each submittal.

• Keep the review process practical using common sense. Don't hold up the job because one useless test report was missing from the submittal package. You can send back the package with a note "approved with comments". The comments may be to submit that report before the work starts.

• As mentioned earlier, keep a good log that is visible daily.

• Create your checklist for all needed submittals to make sure nothing is left behind.

• Same comments about the squeaky wheel. Follow up, follow up, and follow up.

• Maintain a log to track the needed submittals and their status. You don't want to be responsible for delaying the job because the submittal was not reviewed in a timely manner.

One thing to remember about logs, which is extremely important: Enter information in log *the moment* they occur. If you postpone logging information till Friday, chances are your logs will be out of date because of the new fifty problems that come up the next day. For example, you get a submittal, enter that date in the log and then process the submittal.

Mobilizing and Starting the Construction Project

At this point, the owner has issued the general contractor a Notice to Proceed (NTP). The first step is to mobilize on site and start the project. This chapter will be dealing with this phase.

The starting and the closing of a project are usually the tense parts of a project. I am not saying that the rest of the duration is problem-free. I know it isn't, but these two phases take more than special attention.

What is involved in mobilization/starting the work?

If you are a construction manager for a general contractor, here is a list of the tasks at hand:

- Setting up the temporary fence to enclose the construction area.
- The site temporary office
  - Connecting temp power to the site office.
  - Connecting plumbing and drainage for the site office.
  - Setting up Internet and phones.
  - Site office furniture and supplies.
- Temporary toilets
- Placing project sign
- Setting up the water service for the construction activities.
- Setting up electric power for the construction activities.
- Pulling permits to start work.
- Move-in equipment for the first trades.
- Storage bins
- Trash bins
- Set up construction lights.
- Set up pedestrian signs and traffic control.
- Issue the subcontract agreements.
- Set up your logs.
- Complete the immediate required documents like the baseline schedule.

Before we get into each one of the above, I would like to point out that the owner's construction managers need to think about the site logistics and the layout of where to place the above items before advertising the project. In fact, one of the plans should be for site logistics.

The reasons for this advance thinking are obvious. You need to pre-plan a project so it flows smoothly. The site logistics affect the bid price, so you don't want to leave that unaddressed. You do not want to award a job to the contractor who then will claim that they didn't bid the project to have a staging area fifty feet away from the job site, or claim that the specs did not specify a project sign, an 8' high chain-link fence, or a 48' site office. All this will be change order material.

If you think I am exaggerating, here's an example, which you may think ridiculous, but it really happened. One of the owners I worked with had in specifications: "The contractor shall provide a copy machine to the site office." One smart jack provided a copy machine. No one specified that it should be working. The owner modified their specification to say, "The contractor shall provide a functional copy machine to the site office." I'm not joking, this really happened.

Remember the five Ps of construction: Poper Planning Prevents Poor Performance . Now, let's get into each of the above items and talk a bit about what is involved.

Setting Up the Temporary Fence to Enclose the Construction Area
• Review the specification for the requirements regarding the fence. Is it 6' or 8'? Is a tarp required or not? Etc.

• Coordinate with the owner's representative for the setup.

• Plan this well based on the scope of work so you don't have to pay for moving the fence repeatedly.

The Site Temporary Office
　　○ Connecting temp power to the site office.
　　○ Connecting plumbing and drainage for the site office.
　　○ Setting up internet and phones.
　　○ Site office furniture and supplies.

• Review the specifications for the requirements for the site office. Size, number of rooms, etc.

• Plan the location well in light of the scope of work so you don't pay for moving it twice.

• Follow the staging requirements in the contract documents. If none exist, coordinate with the owner's representative.

• Temporary power
　　○ Set it up with the local power company.
　　○ Plan the location of the power poles to provide power to the work areas and minimize moving them during construction.

　　○ Check the contract documents; who pays for this power?

- Plumbing
  - If the temp office has a restroom, it will need plumbing work, water supply, and drainage.
  - You can either connect drainage temporarily to an existing sewer line, or you can get a portable drain box that takes in the sewage, which has to be cleaned regularly. Check your budget to decide what to do.
- Phones and Internet
  - These days, you can't run a site without internet and access to email. The cheapest way is to get these wireless devices to provide internet service.
  - Or you can pay for a phone/Internet package.
  - Set up you phone/fax service.
  - Realistically, site staff will use their cell phones.
- Site office furniture and supplies
  - Some specifications list exactly the office furniture that is required.
  - If not, just get what you need to run the project.
  - The standard items are desks, filing cabinets, copy/scan machine, chairs, water dispenser, etc.

Temporary Toilets

Check your local regulations about the ratio of temp toilets to the number of workers. For example, 1 to each 10 workers.

Select the locations to be practical to workers and the truck that needs to access them to pump out the sewage.

Placing Project Sign

The project sign is important to owners to tell the public about what is being built and who are the main parties. It can also include the names of local politicians that get credit for the project.

The contract documents typically specify the sign's size, wording, material, etc.

Submit the details to the owner for approval and discuss the location if not already specified.

Setting Up the Water Service for the Construction Activities

The general contractor and its subcontractors will need water for their work. If the site is large, then a network of water supply lines has to be strategically designed to serve the work and be out of the way.

There are many ways to get water. The contractor may get a temporary water meter connected to the main line, hook up a meter to an existing fire hydrant, etc.

Make sure you check the local water company for the relevant regulations. One last thing, don't forget to have more money in your budget for this item.

Setting Up Electric Power for the Construction Activities

Similar to water, the construction trades will need temporary electric power. There are companies that specialize in providing this type of service. They will deal with the local power company for the paperwork and the needed work.

The temporary site office will need electric power to operate the office equipment and the HVAC on those miserably hot days of the year.

Typically, an existing power source is identified; then power poles are installed throughout the site so subcontractors can get power for their work. On some jobs, generators will be provided.

Again, plan the locations of the poles to be out of the way and moved the least amount of times to reduce cost.

Pulling Permits to Start Work

This is one of those cost items that a lot of general contractors miss when preparing their bids, especially in the rush of public works projects. In the private sector, this cost item is listed as an exclusion because these permit numbers are not known until the contractor actually goes to the city to pull the permit in most cases.

In the public sector, it is a mix between the public owner picking up the cost of all permits, or some, or none.

If you happen to be the contractor, make sure you look for this item in the bidding documents; otherwise, you will end up paying out of pocket, in a lump-sum bid.

If you are the contractor, you also need to research the permit requirements and time elements so you plan your budget and sche-dule.

This is obviously one of the first sets of tasks to do when the job starts.

Move In Equipment for the First Trades

There is nothing more enjoyable on a job than being ahead of schedule. As you are planning the start of your project, list the first trades to start within the first month, connect with these subcontractors, and have them move their equipment and materials so they are ready.

It takes a while to set up and sign up all the subcontractors on a job. My best advice to you is to get those early trades going while you deal with the rest of the trades.

Storage Bins

Typically, each subcontractor gets his own storage bin on site to store his material. So if you have forty-five subs on the job, you'll need to plan the logistics of where they go without blocking access and the work areas.

In some cases, the general contractor provides storage for a cost or part of their negotiation with the subs.

Security becomes an issue. Typically, the general contractor provides security to the site as part of the general conditions cost. Other contractors leave vandalism incidents to be handled by each sub's insurance.

It is also a cost item that is missed a lot.

Trash Bins

Who gets what and who pays for what is governed by contracts between the parties. So first thing you do is check that pays for the trash bins. Because once a trash bin is on site, it becomes the dumping site for all subcontractors, the neighbors, and their ugly cousins.

This is also a major cost as bins will get filled very quickly. So we are talking rental costs and dumping costs that can eat up your budget. Here are the various scenarios:

• The general contractor provides trash bins for all subcontractors to use.

• Each subcontractor gets its own bin.

• Some subcontractors get their bins, and some use the general's .

• There are specialty bins, like for concrete that the concrete subcontractor provides, or a hazardous materials bin that the hazmat subcontractor provides.

Keeping the site clean is a safety issue and a standard requirement of owners. So please make sure that you review this task when you prepare your budget.

The other issue is planning the location of the bins on site to serve work efficiency and make it easier for the haul-off truck to reach the bins.

Set Up Construction Lights

This is especially important when you have interior work going on where there is no power connected yet. Not having lights in all areas of work is a serious safety issue.

If you are the contractor, make sure you budget for this and plan the network for these temporary lights.

Set Up Pedestrian Signs and Traffic Control

There are several site situations where traffic and pedestrian controls will be needed. There is a cost attached to each measure that can vary from a few hundred dollars to thousands. This is another common cost item that I have seen a lot of contractors miss in their bids. Let's explore the various scenarios.

A Modernization Project of an Existing Building

In this type of project, the contractor will be doing mostly internal renovation work where occupants of the building may still be in there using parts of the building. In this case, the areas under construction have to be barricaded with signs directing pedestrians how to go around it, warning signs, and other signs that may be required in the contract documents.

This project may be done in phases, and the barricades and signs have to be moved. Fire exits and accessible pathways have to be considered.

Check the contract documents regarding this issue for specific requirement.

A Project Where the Public Sidewalk May Be Impacted

The work may include exterior renovation that will impact pedestrian traffic along the sidewalk next to the building or the street along the building.

In this case, depending on the local regulations, the contractor may need to prepare a traffic control plan that has to be approved by the local government agency. This plan will include the details about blocking the street, directing traffic, placing barricades, the time of work, the number of workers, etc. There are companies that specialize in this type of service.

There may be a need to place covered scaffoldings to avoid interrupting traffic along the sidewalk and to protect people from falling debris from above.

Public Sidewalk Illustration

Public Sidewalk Photo

This can be a substantial cost item. I keep saying this because I have seen many contractors ignore allocating money for this task in their budget and end up losing money.

Issue the Subcontract Agreements

It takes a while after the general contractor has been awarded a new project to have all the subcontractors' agreements negotiated and signed. There are several reasons for this:

• The general contractor bids a project using numbers from the subcontractors or their own estimates.

• The subcontractor's number may be above budget and needs to be negotiated.

• The subcontractor's proposal may not have included the complete scope, or there are some exclusions that the general contractor does not want to pay for out of their budget.

• Most general contractors have template contract agreement forms. Some subcontractors may have some issues with some of the clauses that have to be negotiated.

There can be more reasons, but you get the point. So in order to start the project and not lose time, it is wise for the general contractor to focus on the immediate trades and get their subcontract agreements taken care of. This way, the job can start while finalizing the rest of the contracts.

The other priority contracts should be for the trades that have long lead time, items like structural steel, doors, hardware, storefronts, etc.

If you are working as a construction manager for the contractor, the first thing you do is prepare a log of all the trades on the job to use as a checklist to make sure that all trades have been covered.

Set Up Your Logs

I am a big fan of logs. In my opinion, logs provide a powerful tool to track tasks, that is, change orders, RFIs, RFPs, submittals, delay notices, inspector's noncompliance notices, etc.

If you want your logs to work for you and stay current, here is the trick:

Note your log entry the moment the tasks take place. If you postpone that, your logs will always be incomplete and off.

For example, you send an RFI, open the log, type it in, and then process the paper. You receive an answer to the RFI, open the log, type it in, and then process the paper. You send a submittal, open the log, type it in, and then process it, etc.

You will get very busy with new problems every day. If you join the "I'll update the logs on Friday" club, here is what will happen. Friday will come, you are tired from the week, you look at your tray, and there is a thick five-inch stack of papers waiting to be logged, and you feel like you need to relax. Trust me, I learned the above advice the hard way.

Let's go back to logs. Make a list of tasks to be tracked. Form a log for each task and have them ready for entries. Do that at the job start before it gets hectic so that entries are done on time.

I will address the topic of logs later in this book when I talk about construction documentation. Oh! One more thing to remember about logs:

Logs that are not visible or easily accessible are useless.

Logs are meant to track. You need to look at them daily to know what is still open and what needs a follow-up call, not to mention the dates, which are very important. So make them easily accessible and visible to use them effectively.

Complete the Immediate Required Documents, Like the Baseline Schedule

The contract documents often list required documents to be submitted by the contractor at the start of the project. That includes the following:

- Baseline schedule
- Schedule of values
- Site logistics plan
- A list of subcontractors and vendors
- Insurance certificates
- Etc.

Check the contract documents for these requirements. *Form a log* so you can check them off, and track them so you can meet your contractual requirements.

If you are working as the construction manager for the owner, you need to make a similar log to make sure that the contractor has submitted all the needed documents.

In conclusion, as you can see, starting a job is not simply "Let's get the truck and go to site." Typically, starting and closing a project are intense occasions and need special attention.

Meetings

As a construction manager, you will be part of several meetings through the life of the project.

We talked in previous chapters about some of the meetings that take place on a construction project, namely the pre-construction meeting and the pre-bid meetings. Take a look at the list below. It will give you an idea about the typical types of meetings that will take place. You may be a participant, or you may be the person putting together and leading the meeting.

If you are a construction manager representing an owner:

- Program planning meetings
- Scope planning meetings
- Design review meetings
- Internal staff meetings
- Meeting with the owner's team
- Meeting with the owner's upper management team.
- Meeting with local community members.
- Board meetings
- Weekly construction progress meetings with contractors.
- Claim meetings
- Schedule review meetings
- Etc.

If you happen to be a construction manager for a general contractor:

- In-house staff meetings
- Weekly site meetings with the owner.
- Schedule review meetings
- Job start meeting
- Subcontractors coordination meetings
- Claim negotiation meetings
- Etc.

The above list is self-explanatory about the topics of each of the meeting. The objective of this section of the chapter is to talk about the dynamics of meetings, I hope I can give you some pointers about making these meetings successful.

Before we go there, I would like to mention one obvious comment, which is that you need to gain and improve on speaking skills, organizational skills, and understanding of human nature. Remember the main theme of this book, which is the technical portion, is the easy part; human nature and its imperfections are the real challenge. So whether you are a participant or the leader of the meeting, dealing with the human element is crucial to the success of the meeting.

Here are some tips to have successful meetings:

- Plan the sequence of topics to flow logically.
- Don't invite people who have nothing to do with the topics. They will feel that you wasted their time.
- Don't make it too long that people lose focus and interest.
- Plan it so everyone feels they were important to the meeting, that they contributed, and that they leave the meeting feeling positive.
- Give credit to people who earned it.
- Send the agenda ahead of time so people are prepared.
- Don't embarrass people in front of everybody.
- Think about the location to be well prepared.

Real-Life Gauge

The advantage of having regular meetings is that it keeps people on their toes and under pressure to perform. When meetings get cancelled, the tasks drag as people have other projects to deal with. I have seen it happen many times. People think, "I better get this task done before the next meeting."

The other thing I don't understand about meetings is attendees see each other daily, but wait for the meeting to throw their bomb at you and embarrass you in front of everybody. And some love the moment that they can point out your mistake or what you said was wrong. Why the heck didn't they mention something before the meeting?! Why did you have to wait until we are in front of all these people?!

Have you ever been in meetings where the topics are so unrelated to you, and you wonder when is this nightmare going to end? Remember that next time you plan a meeting. Plan your topics and choose the attendees.

Make it positive. There were times where I told people before the meeting what was going to be asked of them during the meeting. Is that cheating? No, I think that serves all purposes; it gets the tasks done and has the person involved feeling positive and, therefore, more productive.

Another important aspect is the documentation of the meeting. Someone should take meeting minutes and later distribute them to the attendees. Personally, I prefer to write the minutes instead of having to review someone else's writing, which may not be objective. Once meeting minutes have been published without any protest, they become documented facts that have value should the project go into claims. So be careful what is written down.

Request for Information (RFI)

We talked above about the various phases that a project goes through, from scope planning to design and construction. There will be many questions that will come up from various sources that the architect needs to address. The official method of documenting questions and answers is by the use of an RFI form.

This form has a section for the question and another for the owner/architect's answer. There are several forms for an RFI with different set of information to be shown on the form. Please see below a sample form and check the points.

The main components of an RFI are the following:

• Dates. It is so often that I see RFIs being sent by saving the last RFI without changing the date. The date that the question was sent and answered has a great impact and significance on delay issues. If a certain RFI was answered late and impacted the critical path of a project, causing delays, then these dates become crucial evidence.

• Who is the RFI going to, and who answered the question, and who approved the answer? These names are important to be right because of liability allocation in case the question or answer results in disputes.

• The body of the question. Please make sure that the question is valid and clearly explains to the receiver what you need. There is nothing more irritating than getting a question whose answer is clear on the plans. Now, time has been wasted processing and logging this document.

• The body of the response. I have seen several responses that did not answer the question or tried to play smart games to avoid admitting that the plans missed information. Needless to say that this only generates animosity and claims. If you are an owner's construction manager, please screen the architect's answer before sending it to the contractor.

<table>
<tr><td colspan="4" align="center">REQUEST FOR INFORMATION FORM</td></tr>
<tr><td colspan="2" align="center">SUBMITTED FROM</td><td colspan="2" align="center">SUBMITTED TO</td></tr>
<tr><td>COMPANY:</td><td>ABC Construction, Inc</td><td>COMPANY:</td><td>Architect, Inc</td></tr>
<tr><td>ADDRESS:</td><td>555 Main Street
USA</td><td>ADDRESS:</td><td>555 Main St
USA</td></tr>
<tr><td>PHONE:</td><td>555-555-5555 FAX: 555-555-5555</td><td>PHONE:</td><td>555-555-5555 FAX: 555-555-5555</td></tr>
<tr><td>EMAIL:</td><td>Joe@xyz.com</td><td>EMAIL:</td><td>Suzan@dcf.com</td></tr>
</table>

PROJECT:	Name of Project	ARCHITECT'S PROJECT #:
ARCHITECT:	Architect's Firm	BID PACKAGE #:
ATTN:	Architect's Name	RFI #: **45**
FAX:	555-555-5555	DATE REQUESTED: 10/20/2015

INFORMATION REQUESTED:

Please clarify the footing's depth in detail 5/S1.1

DRAWING TO REFERENCE: S1.1	*SPECIFICATION SECTION:* 33000

CONTRACTOR'S RECOMMENDATIONS:

Recommend 24"

IMPACT CONTRACT TIME: YES / NO	RESPONSE NEEDED BY: **11/2/2015**
IMPACT CONTRACT COST: YES / NO	

SUBMITTER'S SIGNATURE: ________________   DATE: _______

ARCHITECT/ ENGINEER'S RESPONSE:

Use 36"

ARCHITECT'S SIGNATURE: ________________   DATE: _______

(Provide attachment if additional space is needed in the above sections)

Real-Life Gauge

Keep in mind the main theme in this book: Building a trusting relationship between the contractor and owner is key to the success of the project.

You may ask, what does this have to do with an RFI? Well, there are some contractors who use the RFIs to create false situations to generate change orders or delays. So if the wording of the question conveys that impression, the contractor is impacting the trust relationship with the owner. By the same token, I have come across RFI responses that owners and architect try to pin a missed item on the contractor to avoid admitting to their mistake. That also impacts the trust and relationship issue.

Here is an example of an RFI that is a problem:

Question: During excavation, we exposed unforeseen underground plumbing lines that are in conflict with the new ramp.

Answer: Now that they have been uncovered, they are no longer unforeseen and therefore are part of your means and methods.

If you are a contractor reading this, I am sure you would have screamed already. These are the type of word games that destroy any trust on the job and turn small issues into big problems. By the same token, I have seen questions from contractors asking for a change order on work that is clearly specified.

What I am trying to say here, whether you are representing a contractor or an owner, just keep it straight and fair. Owners and contractors need each other's help to finish the job, and no one wants to pay for what is not theirs to pay for.

The wording of the questions and answers play a big role in creating the trust that is needed for the team to act like a team. Like it or not, there is a lot of stereotyping in this industry about how contractors view owners and architects and vice-versa . The fact is stereotyping doesn't come from a vacuum. The behavior of these parties has earned them these perceptions.

You know how they say that first impressions last forever? Take the early stages of the project as that first impression and convey a feeling of fair play and honesty so that bigger problems down the line get resolved smoothly. Everyone benefits from this.

Best Practices

If you are a construction manager for a contractor:

- Don't send a ridiculous question or one where the answer is clear on the plans.
- Log the RFI as soon as you generate the RFI. Make sure you log the date it was sent and a description of the question.
- If the issue may result in added cost and a time extension, please note that, which is providing a notice to the owner. This is very important.
- Process the RFI the same day you get the question from your subcontractor.
- Include what subcontractors need to know about this question.
- Log in the date you receive the response the moment you get the response.
- Review the response. If the response did not give the needed information, you need to inform the owner immediately.
- Forward the response to the relevant subcontractors immediately.
- If the response constitutes a change in cost or time, notify the owner immediately.

If you happen to be a construction manager for an owner:

- Log in the RFI that you receive from the contractor the moment you get it.
- Send the RFI to your architect and log the date you sent it.
- Follow up with the architect so the response doesn't take longer than what the contract specifies. A late response will put the owner in a position of liability.
- When you get the architect's response, check the answer. If the response did not address the question, evaded the question, or presented a suggestion that may result in an unnecessary additional cost, discuss that with the architect before sending the response to the contractor.
- Don't send the contractor the wrong response, as this shows a lack of management on your part.

As mentioned earlier, bidding a private-sector project is different compared to public works. We also covered that in previous chapters.

What I would like to add here is the importance of communicating with the subcontractor regarding the bid versus taking in whatever number they provide and using it in the bid. If they should miss an item, the owner will hold the general contractor accountable while the subcontractor will be demanding that cost because it was clearly excluded in their bid.

When there is a budget to meet, and you need the subcontractor to lower the cost, this is a good example where the relationship factor plays a big role.

Just like anything in life, nothing is free of problems and issues to think about. Here are some to be aware of:

What Paperwork Do You Need When Dealing with Subcontractors?

When you first hire a subcontractor,

- Make sure their license is current and the right license for the trade;
- Get the required insurance certificates, naming the general contractor and/or the owner as additionally insured; and
- Some projects' performance and payment bonds are requested.

What You Need to Know When Paying a Subcontractor

Subcontractors have laborers, suppliers, and second-tier subcontractors to pay. A second-tier subcontractor is a firm hired by the subcontractor. It is a called second tier because the subcontractor has a prime contract with the general contractor or the owner.

The problem is when a subcontractor gets paid, and they don't pay the above-listed parties. What happens then is those people may go after the general contractor and/or the owner. Second-tier subcontractors or vendors may file stop notices of lien to the private property. In public works, the general contractor may be liable for the laborers that the subcontractor did not pay.

So how do you protect yourself if you represent an owner or a general contractor? Before I list some options, let me explain about lien releases.

• *Conditional Release for Progress Payment.* Let's say a subcontractor sends an invoice for charges up to September 30 for $50,000, and this is not the final invoice of the job. A conditional release for progress payment is a document saying that they are owed $50,000 as of September 30 and that if this money was received, they will have no rights to file any liens for payments up to September 30.

• *Conditional Release for Final Payment.* The same principle as above, but for the final payment. That documents that this will be all that is owed at this time.

• *Unconditional Release for Progress Payment.* When the subcontractor receives the $50,000 for the September progress payment, this document confirms that the subcontractor received this amount and they waived their rights to file a lien or a stop notice up for money owed up to September 30.

• *Unconditional Release for Final Payment.* Same principle, but for the final payment.

Having said all that, let's go back to measures the general contractor or the owner can take to protect themselves.

• Owners or general contractors should not process payments until all releases are submitted.

• Owners and general contractors can pay joint checks to the subcontractor and their subs and vendors.

• In public works, the general contractor may be on the hook if the subcontractor didn't pay their laborers. That is why all certified payroll reports have to be current.

Finally, remember that the subcontractors hold most of the risk, and they are the ones building your projects. It is in your interest to make sure that they succeed and don't lose money. Eventually, their failure will be yours.

Cost Control

What are the three most important elements of a construction project?

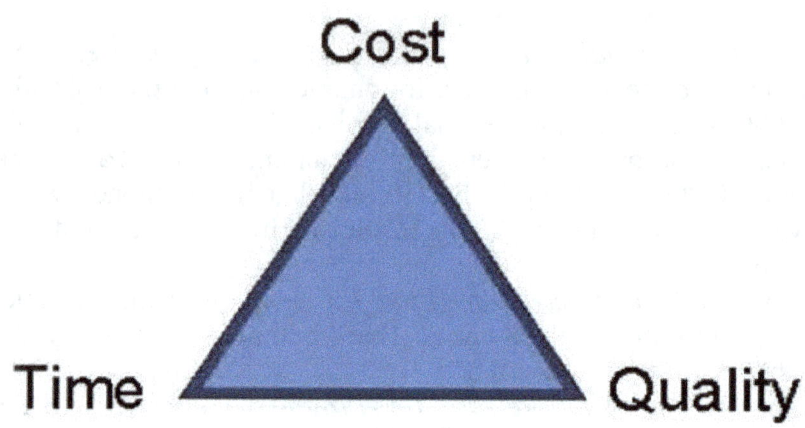

If all three are satisfied, the project is a success. In this section, we will address cost control.

In a previous chapter, we described the various phases of a construction project. Cost control starts and continues to be a factor, from planning up to completion of the project.

Planning Phase

This is the phase where you are determining the scope. That means you are setting up the potential cost, which is directly related to the planned scope. Cost control is needed.

Design Phase

If the design is not being monitored and you end up with a larger scope or costly details that could be built cheaper, then the budget will be busted. Interim estimates are needed as the design is being developed. Cost control is needed!

If the owner doesn't do constructability reviews to foresee potential claims, change orders, and flag errors and omissions, the change orders and claims will eat up the budget. Cost control is needed!

Construction Phase

This is the most critical phase where cost can spiral out of control if the project is well managed. Cost control is needed!

Okay, now you are probably tired of hearing "cost control is needed." Well, let's get into some tips and best practices on how to do that.

Needless to say, there are so many choices when it comes to construction management software that records and tracks costs. I'll leave that topic for someone conversant about software and computers.

• A good estimator is priceless. From the early planning stage of the project, start with an estimate for the various phases of the project. Once approved, use that as the master budget that we need to work from and try to save moving forward.

• Establish a system of recording actual costs for each of the cost line items. This system has to allow you to enter estimated and projected costs as well to compare against the master budget.

• If the projected numbers increase the bottom line of the overall master budget, then look for future activities and see where you can cut costs to go back to budget.

• Refer to tips in previous chapters about how best to manage the various players of the project. Each one of them has their own budgets and methods of operating. Make sure that your decisions and management style are not causing them additional costs. For example, we covered before what not to do when working with architects, like asking for a lot of changes in the middle of the design development.

• As the design is being developed, run interim estimates to make sure the design is not expanding out of control.

• Run a final estimate before you advertise the project for bids.

• Do your due-diligence constructability reviews before advertising the project.

• The construction phase is the most critical phase in cost control. This phase is the reality check of all the pre-construction mistakes and the problems after the start of construction. Here are typical examples where cost may spiral out of control:

° The bid from the contractor is very low, and the contractor can't finish the job. Completion costs are usually much more than the remaining balance of the contract amount, not to mention possible legal fees. So what should you do?

- Evaluate the bids before award.
- Don't overpay the contractor.
- Be a part of the solution to problems and not complicate matters that may drive contractors to lose money, which may lead them to not be able to finish the job.
- Apply proper project manager practices, including understanding the terms and conditions of the contract, resolving issues before they escalate, and being fair.
  ○ Change Orders. This is a sour topic to many in the industry as they may invite conflict. If you happen to be the owner's representative, you will have to do the following:

(a) Be fair in assessing the validity and costs of the change orders.

(b) Look for alternatives that may eliminate the need for the change order or reduce its cost.

(c) Maintain a good log that tracks and projects possible future change orders. That will help you take decisions to control the cost. For example, if you project that the cost will go beyond budget, you may want to meet with the owner and the architect to discuss adjustments to the remaining scope.

○ Owner's Requests. It is not unusual for owners to ask for additional work or modifications of the current scope during construction. What you should do is have a spreadsheet or some software that can record the current cost and the projected completion costs. So when the owner asks for additional work, you can plug in an estimate of this new request to advise the owner as to the impact on the budget so proper decisions are taken. I have seen projects where owners kept directing contractors to do additional work without tracking, only to get surprised at the end with a huge cost overrun. What do you think happens when the owner doesn't have enough to cover the final cost when the contractor has already done the work? Hello, lawyers!

○ Incompetent Project Management. Need I say more? When the construction manager doesn't manage properly with bad paperwork and decision-making or unfair handling of issues, the project will probably go legal, and there goes the budget. This book should be treated as a valuable resource for managing the various phases of the project, which is the key to cost control.

The most important tool for cost control is to use a system that can record the original budget, track the current costs up to date, and the ability to project potential cost to complete the project. That could range from a simple spreadsheet to sophisticated software. The tasks are the same.

Documentation, Communication

It would be nice to think that construction is only about building things. No! That would be too easy. We talked a lot about the human factor in projects, but there is also the critical task of proper documentation.

As a project manager, a lot of my time goes into documentation, emails, letters, etc. Why is documentation so critical?

Let's think about it from the end, going backward. If you get into litigation or claims, whether during the project or three years after completion, what the lawyers and others will look at to argue their cases are the project documents.

- What document can be used to prove whose case?
- Content spinning will get very creative.
- If you are present in the room when the lawyers from both sides are arguing, you may wonder if they are talking about the same project that you managed.
- The other side may forget how nice and helpful you were, and the fact that you never documented an issue may be the key to their case.

The good news is that if you do your documentation well, that may decrease the chances of getting into litigation. The other side will see the strength of your case because of your documentation, which may deter them from pursuing the case and settle.

Forget litigation for now; documentation is also very critical to establish clarity of information, document decisions, changes from the plans, etc.

As a construction manager, whether you are the owner's representative or the contractor's, you have to have a good system for documentation, logging, document control, and an eye for potential liabilities because…*every document that you write can be used against you in court.*

Please remember this golden rule before you issue any document, letter, or email. Speaking of emails somehow, people forget that emails are documents that can be presented as evidence, and they type like they are chatting on the internet —so carelessly and without any consideration for the consequences.

The contract documents are the basis of every debate and issue on the project. If it is not in writing, it didn't happen. You want to avoid the selective amnesia situations and getting into "Don't you remember when we were standing by that tree and we agreed to use blue paint for the west wall?" or "I thought that we agreed to deliver the cabinets tomorrow," etc.

Okay! Enough blabbering about documentation. Let's get into it.

We talked about the various phases of a project. Each phase has its own set of documents that the CM has to carefully handle. You are the consultant and may be held liable if your client gets in trouble because of your negligence in not properly documenting an issue or filing the right notice on time, etc.

Here is a list of typically needed documentation.

Typical Documents – Planning Phase
• Scope of work. After several meetings with the owner and stakeholders, you will need to document the agreed scope before moving forward with the design. Make sure you get the signoff from the right people. Last thing you want is spend money on design to get something like "We never said we needed a library in the building."

• Meeting minutes documenting discussions and decisions. Have a sign-in sheet to document who attended the meeting. Then issue meeting minutes to all relevant parties to document the discussion.

• Budget and estimating. After the scope is done, run an estimate or create a budget that the client will sign off on before you move into design.

• Schedule. Develop the master schedule for the project and get signed off.

• Emails and correspondence (don't treat them lightly, especially emails). What you write may be used against you. The main idea is to document decisions and conversations for any future disputes.

Topical Documents – Design Phase (See Notes Above)
- Scope of work.
- Contracts with designers and consultants.
- Interim plans and specs.
- Design review comments.
- Estimates.
- Emails and correspondence.
- Meeting minutes.
- Bid documents. These better be complete, or the owner will be subject to change orders from omissions. The bid documents have to include all the information and instructions that bidders need to bid the project.

Typical Documents – Bid/Award Phase
- Advertising. Check the relevant legal requirements of what information should go on the advertisement, especially for public works projects.
- Instructions to Bidders. This outlines the documents and procedures for bidding the project.
- Bid Forms. The forms that bidders have to submit with their bids.
- RFPs, RFQs (Request for Proposals and Request for Qualifications)
- Bid Package. The package that is given to bidder containing the above information.
- Bidder's Questions. After advertising the project, the bidders may have questions. The instructions will tell them who to send it to. Make sure that all bidders get a copy of all questions and answers so that all bidders are using the same information. In public works, if this doesn't happen, it may be grounds for protests and bid cancellation.
- Addenda. Any changes from the original bid documents have to be issued in writing to all bidders.

Typical Documents – Construction Phase
- Notice of Intent to Award. When the successful bidder is selected, the owner issues this notice to inform the contractor on

their intent to award. Perhaps the contractor can start preparing their documents and their subcontractors.

- Notice of Award. The actual notice.
- Notice to Proceed (NTP). The start of the clock for the project's duration.
- Job Start Meeting Agenda. We covered that in previous chapters. It is the pre-construction meeting before the start of construction.
- Meeting Minutes
- Schedule of Values
- All Correspondence
- RFC (Request for Clarification)
- RFP (Request for Proposal)
- COP (Change Order Proposal)
- CD (Construction Directive)
- CO (Change Order)
- Delay Notices. This is a very sensitive issue. Any cause of delay has to be notified in a timely manner, or the claimant may lose their delay claim.
- Inspections Requests. It is typical for the contractor to file a written inspection request for various items, like check rebar before pouring concrete etc. Make sure you read the contract to check if there are any required timelines for inspection requests. Some contracts may ask for twenty-four hours, some forty-eight hours.
- Inspection Inquiries. These are questions raised to the architect by the inspector. Similar to the RFCs raised by the contractor.
- Transmittals Letters. Make it a habit to issue a transmittal letter every time you are delivering an item to another party. For example, submittals, pay requests, etc. Make two copies and have the receiver sign for delivery.
- Notice of Default. If one party defaults on a contract obligation, a notice has to be given. The idea of a notice is to point the issue in a timely manner and give notice to the other side to give them a fair opportunity to mitigate the issue.
- Beneficial Occupancy Certificate. An owner has the right to occupy a part or all the building before the final completion of the project. Each contract has a different process for this.
- Certificate of Substantial Completion. There are many definitions of what constitutes substantial completion. In general, it is the

point in time when the building is ready for its intended use pending minor corrections. By the way, forgetting a whole elevator is not a minor correction. I'm talking patching a wall corner, fixing some paint, etc. The significance of this point in time is that it typically stops the clock for assessing liquidated damages for delays. Different contracts vary on this issue.

• Punch List. This is the list of minor corrections given to the contractor after substantial completion. Typically, contractors are given thirty days to complete the corrections.

• Notice of Completion. This is a document that the owner files when the contractor completes all contract requirements, including documents.

• Certified Payroll. In public works, contractors have to pay prevailing wages. They have to file these forms weekly, listing all the workers during that week and how much they were paid per hour and total for the week. That is how owners monitor and ensure compliance with prevailing wages requirements.

• Pay Requests. This is the contractor's invoice. They typically file that once a month. Different contracts vary on the requirements for this task.

• Daily Reports. The contractors have to fill out a daily report that they file at their office and give a copy to the owner. This documents what took place on site each day, the weather, number of workers, subcontractors, issues, etc.

Typical Documents – Closeout Phase

• Warranties. The specifications note the warranties needed for each trade and when they become effective and for how many years. The owners use those to call back the contractor to fix problems, like an air-conditioning unit that doesn't start or a roof that leaks, etc. The contract may also dictate the forms to use.

• Operation Manuals. This is like the manual you get when you buy a new DVD player. Examples are HVAC units, book detection system, PA system, etc.

• As-Built Drawings or Record Drawings. During the project, the various subcontractors, like the plumbing subcontractor, have to draw in red line the *actual* locations of the underground pipes that were installed. Or an HVAC subcontractor draws the *actual* location of the HVAC ducts and items above the ceiling. The idea is that the owner keeps a record for future use of where utilities are placed to coordinate future work, design, or maintenance. Typically, a contractor gets an extra set of plans on site that is for the sole use of noting these red lines. This has to stay clean inside the site office. It will be delivered to the owner at the end of the project.

Letter Writing
• Since communication is key to managing any project, writing letters should be done professionally and contractually.
• Take out the ego and think as a third party.
• If you write a nasty letter, wait a day before you decide to send it. So you have time to reflect on what you wrote and reword it.

Here are some key tips to properly write letters to the contractor, architect, and other players on the project.

• Do not deviate from the contract.
• Do not get personal.
• Develop a system to track unanswered topics.
• The following is a sample of the format of a business letter.
• Avoid words like *you, your, I*. For example, "Your superintendent was following your instruction not to install a temporary fence."
• Try this instead: "Please note that a temporary fence has not been installed as of this date per contract documents, section 1026-a.1."
• Always refer to companies, or if you have to name somebody, address them as Mr. Smith or Ms. Smith, IOR.
• The idea is the more you walk away from a personal approach, the document will sound more objective and professional in court.
• How many times did you feel so upset at a contractor that you sat down at your keyboard and wrote a harsh letter?
• Well, do that and let it all out, but don't send it until the next day when you had the chance to calm down and rewrite the wording. Remember, it will be used against you in court.
• This is especially dangerous with emails because when you can click "send," it's all over.
• In the world of claims, only documents matter, and no good deed goes unpunished.
• Always write with an eye for potential claims. Be nice, fair, and helpful, but be protected in case of claims.

See the letter example below. Note the format and the choice of words.

Mr. John Smith XYZ Construction 444 Tree Lane
Los Angeles, CA 90504

Re: Change Order Proposal No. 4
Oz High School, Modernization Project # 3152

Dear Mr. Smith,

Please note that change order proposal (COP) No. 4, submitted by XYZ Construction (XYZ) can't be processed at this time due to lack of proper back up.

We would like to set a meeting with XYZ to discuss COP No. 4 hoping we can get a quick resolution to this matter.

The School District is committed to working with XYZ to ensure timely processing of pending issues to facilitate a successful closure of the above-mentioned project.

Thank you.

Sincerely,

Jane Smith Construction Manager
Oz Unified School District

c.c. Mr. Clark Kent, Senior Project Manager
Ms. Xena the Warrior, Public Relations Consultant

NB: Please invite XYZ's Senior Manager to the meeting .

• Note the format of the sender, addressee, date, subject, space between paragraphs, and no indentation for each paragraph.

• See how I abbreviated Change Order Proposal with COP. You do that in parenthesis when it first occurs, and then you can repeat COP down the letter.

• Note how we address the company name instead of getting personal.

• If you want to name someone, you use *Mr.* or *Ms.* to show respect.

• Note in the last paragraph I said, "The district is committed" versus "I am committed" since this is an issue between companies, not the individuals.

• Note that this documented nicely the fact that the COP was not approved without sounding personal or vindictive.

• *Be careful of what is in writing.* You must always be contractually right; you must never give an instruction that is not backed by the contract or one that can be interpreted like you are abusing your authority.

• Finally, how do you like the people cc'd at the bottom!

Change Orders

A change order is a document that authorizes a contractor to do work that is different from what is shown in the contract documents, which can be adding or eliminating work. The change order identifies the cost change (add or deduct) and any modification to the duration of the project (time extension).

This document is signed by the contractor, owner, architect, and others as specified to be processed. Once processed, the contractor can bill the changed work. Here is how it works.

What Are the Typical Causes for Change Orders?
• Owner's added or deleted scope
• Errors and omissions
• Unforeseen conditions
• Justified compensable delays

Owner's Added or Deleted Scope

The owner has the right to add or remove scope from the original scope of work. At this point, the owner consults with the architect to prepare the documents of the change details. The owner can do one of the following:

1. Send the contractor a RFP asking for a price to decide to move forward or not.
2. Issue a directive to the contractor to proceed and submit their price for negotiation.
3. Direct the contractor to proceed on a time and material basis (T&M). This is when the contractor records daily the labor, material, and equipment and submits the T&M ticket to the owner's team for signature. At the completion of work, collect all tickets and issue a change order for the amount and possible time extension.

Errors and Omissions

It is as it sounds; wrong (errors) or missing information on the plans (omissions). For example, the plan does not show how to connect the steel beam with the CMU wall (omission), or the plans shows a dimension that doesn't work (error).

In this case, the contractor sends a RFCasking about the issue. The architect responds. If the response asks for added work, the contractor sends a notice to the owner, who decides to follow one of the three options in the above section leading to a change order.

Unforeseen Conditions

As the contractor is excavating or demolishing a ceiling, they uncover plumbing lines or some condition that is not shown in the contractor documents. The contractor should send an immediate notice to the owner about the uncovered condition asking for direction. The owner consults with his/her team and directs the contractor. Since this is an unforeseen condition, the contractor could not have included this in his/her bid. The owner can take any of the three options listed in the above section leading to a change order.

Pricing a Change Order Proposal

The obvious charges are the labor, material, and equipment that will be incurred to execute the change order work and the allowed markups.

Each contract is different. The general conditions specify

what can be charged and what is not allowed to be charged, like drill bits or main office staff, etc. This is important to know whether you are preparing or reviewing a change order proposal.

The other important element is the backup documents to the charged costs. For example, if there is a charge for light fixtures ($65,000, for example), the change order document should include a quote or an invoice from the supplier to prove the amount.

If the change order proposal includes time extension, then a time impact analysis should be included to justify the added days and possible compensation. Please refer to the chapter in this book regarding delay analysis.

Below is a sample form of a change order proposal. Note the comments explaining the various fields on the form.

Real-Life Gauge

Change orders are probably the most sensitive elements of a project. They have the potential to bring a lot of aggravation and an adversarial environment.

Have you ever called a plumber to your house to fix something? The plumber looks at the work and gives you an estimate of $350. You think, "Oh well, I didn't want to pay this much, but we have to." You authorize the plumber to proceed with the work. You're standing there watching your bathroom become a mess. Then the plumber calls you and says, "That valve there is busted, and it needs to be changed. I can't complete my work without changing it first. That's an extra of $750." You probably think, "When you looked at the job, didn't you know that the valve needed to be changed in order for you to do your work?"

How are you feeling right now about your plumber? Cheated? Taken advantage of? Or do you say, "He's right, I need to pay the extra with a smile"?

Exactly. That is why change orders are so irritating to owners.

CHANGE ORDER Proposal (COP)

C.O.P. No.:	25
Ref. No.:	RFP 15
Date:	Date
Project Number:	97-34590
Contract Number:	54978

Project Name: Main Street Library
Project Type: New Construction of a Library
To: The Owner's Representative
From: ABC Construction, Inc.
(Contractor)

Description of Work: Addition of a new office inside the Meeting room

A. Subcontractor Cost of the Work

Subcontractor 1	$ 5,000.00	This is the section for the costs by the Subcontractors
Subcontractor 2	$ 6,000.00	
Subcontractor 3	$ 15,000.00	
	$ -	
	$ -	
	$ -	
	$ -	**Subtotal A:** $26,000.00

B. Contractor Cost of the Work

Payroll Costs (See attached supporting documentation.)	$ 520.00	These are the costs by the General Contractor from the attached cost sheet.
Fringe Benefits at 0.00% of Labor	$ -	
Materials and Equipment (See attached supporting documentation.)	$ 375.00	
Taxes at 8.25% of Material	$ 30.94	
Consultant Costs (See attached supporting documentation.)	$ -	
Supplemental Costs (See attached supporting documentation.)	$ -	**Subtotal B:** $ 925.94

C & D: Contractor's Fee:

5% overhead and profit of Subtotals *A*	**Subtotal C:** $ 1,300.00
15% overhead and profit of Subtotals *B*	**Subtotal D:** $ 138.89

E. Bond Percentage:

Bond at 1.00% of Subtotals A + B + C + D — **Subtotal E:** $ 283.65

Grand Total = (A + B + C + D + E) — **Grand Total:** $28,648.48

☐ The proposal would ☑ Increase ☐ Decrease the Milestones and/or Contract Time by __15__ calendar days.

☐ The proposal does NOT affect the Milestones and/or Contract Time.

☐ The delays (if any) will be determined later.

_______________________   _______________________   Date
Contractor Signature   Title   Date

Follow all applicable procedures and provide all appropriate documentation as required by General Conditions Sections 10, 11 and/or 12.

cc:

PRICING SHEET

C.O.R. No.	25
Ref. No.:	RFP 15
Date:	Date
Project Number:	97-34590
Contract Number:	54978

Project Name: Main Street Library
Project Type: New Construction of a Library
To: The Owner's Representative
From: ABC Construction, Inc.
(Contractor)

Description	Quantity	Units	Labor RATE	Material Cost	Material Extension	Labor Extension	Total
Labor	8	hr	$ 65.00		$ -	$ 520.00	$ 520.00
Material	25	Piece		$ 15.00	$ 375.00	$ -	$ 375.00
					$ -	$ -	$ -
					$ -	$ -	$ -
					$ -	$ -	$ -
					$ -	$ -	$ -
					$ -	$ -	$ -
					$ -	$ -	$ -
					$ -	$ -	$ -
					$ -	$ -	$ -
					$ -	$ -	$ -
					$ -	$ -	$ -
					$ -	$ -	$ -
					$ -	$ -	$ -
					$ -	$ -	$ -
					$ -	$ -	$ -
					$ -	$ -	$ -
					$ -	$ -	$ -
					$ -	$ -	$ -
					$ -	$ -	$ -
					$ -	$ -	$ -
					$ -	$ -	$ -
					$ -	$ -	$ -
					$ -	$ -	$ -
					$ -	$ -	$ -
					$ -	$ -	$ -
					$ -	$ -	$ -
					$ -	$ -	$ -
					$ 375.00	$ 520.00	$ 895.00

The contractor bids the job per what is shown on the plans and specification. Then enter the missing information, site conditions, etc. The contractor asks for a change order. The same above plumber example feelings kick in with the owner and the architect.

The owner didn't budget for this added cost. The architect feels the need to defend his/her error or omission in front of his/her client; the owner and architect may get creative in interpreting some general note to mean that the contractor should have figured out what the architect failed to show on the plans/specs.

By the same token, the situation may be that the contractor is not being straight and found some loophole to argue for a change order that is not clearly a change order.

Either way, this is a source for claims and animosity on the job.

One other myth is that change orders are heaven for contractors. Not always. Take this scenario.

The owner directs the general contractor to proceed with added work and submit a change order concurrently. The general contractor directs the involved subcontractors to do the same. The subs proceed and pay for labor and material. The work is done. Meanwhile, the rest of the subs have not submitted their price to the general contractor, who can't submit the overall COP yet. After four weeks, the COP is submitted. The owner's representative is busy and sits on the COP for another three weeks. After all that time, the GC and the owner's representative sit down to negotiate the COP. The work is done, and the money has been paid. The owner's representative gets smart and starts chopping prices down to negotiate a lower cost. The contractors are anxious to get paid, so they agree to a lower number. By this time, the subs already lost on the interest haven't been paid on time. On top of that, now they have to accept a lower payment.

There is a psychology to change orders. There are some stereotype perceptions in the industry. Some owners and architects think contractors are there to suck every dollar out of the owner, and there are contractors who think that owners will go to great lengths to deny legitimate change orders to cover their mistakes.

There is also another dynamic. Sometimes the COP negotiations turn into an ego competition:

"Oh, you're not going to get this past me."

"Hey guys, I just got out of the meeting, and I cut that COP in half."

The worst one I heard: "I put that contractor out of business."

Or "I had it in my bid, but I got the owner to pay anyway."

A wise team of players on both sides will eliminate these perceptions from day one on the job. Both sides have to demonstrate that fairness and honesty will be the name of the game. Once that trust is built and change orders are assessed and priced fairly, change orders will go more smoothly.

The owner is the one who has to set this fair tone. After overseeing and managing over six hundred projects, I have yet to see a perfect set of plans. Owners have to expect that and set aside some contingency money for change orders and have a qualified construction manager to mitigate the impact and, most importantly, spend time on pre-construction reviews.

Best Practices

If you are a CM for the owner:

• Create a log to track open COPs. Add a column for projected costs of potential COPs to make sure the project is not going above budget. Last thing you want is to get to the end of the job, and there are open COPs without funds to pay for them.
• Process COPs as fast as you can. Don't let them sit.
• Be fair in your assessment of the merit and the cost of submitted COPs.
• Know your contract documents well. You don't want to approve a COP where the charged scope was covered in the contract documents, or you don't want to approve a cost that shouldn't have been charged in the COP.
• From day one on the job, establish a culture of fairness.

If you happen to be a CM for the general contractor:

• Send a timely notice to the owner when you encounter an issue that is a change order matter.
• Maintain a good log to track the status of the COPs. Note down any issue that can be a potential change order.
• The challenge is to get prices from subcontractors fast in a timely manner. Try to submit COPs as fast as possible.

• Charge fairly. Don't blow up your costs. The CM on the other side isn't dumb, and everybody can estimate.

• Start a file for each issue when it happens, and accumulate the relevant documents in that file. These will be your back-up documents in case of a dispute.

Request for Proposal (RFP)

We talked in the previous section about change orders. Well, what if the owner wants to find out how much the cost is before making a decision to authorize proceeding with the added work?

In this case, the owner will issue an RFP. The owner's team will put together the request with the relevant sketches and details. Issue these documents with an RFP to the contractor.

By contract, the contractor would have a certain number of days to submit a price in the form of a change order proposal (COP). The owner and contractor may negotiate the price, and then the owner will decide how to proceed.

If the owner approves the final price, he/she will issue a directive to the contractor to proceed for the agreed amount.

A common mistake among contractors is that they proceed with the work as soon as they receive the RFP. This is a problem because if the owner decides not to do the work, who will pay the contractor now that work is done? In this case, the contractor is at risk of having done work without compensation.

The next page shows an example of an RFP.

REQUEST FOR PROPOSAL *(RFP)*

RFP Number: ______________

School Name: ______________________________   Date: ______________

Project Name: ______________________________   Project No.: ______________

Issued To: ______________________________   Contract No.: ______________
(Contractor)

Please submit an itemized quotation for adjustments, if any, in the Contract Amount, Milestones and/or Contract Time reflecting the proposed Work described herein. The Change Order Proposal must be submitted within 10 days after the receipt of this Request for Proposal (RFP) pursuant to Articles 10.7 through 10.17.

> **THIS IS NOT A CHANGE ORDER OR A DIRECTION TO PROCEED WITH THE WORK DESCRIBED HEREIN.**

Description of Work

Attachments

Issued by: ______________________________   ______________________________
         Name (Printed)

Change Order Proposal is due from the Contractor by:		
	Date	
Owner's Authorized Representative Signature	Name (Printed)	Date

Make sure of the following:

• Log the RFP for tracking.
• Submit the COP in a timely manner.
• Do not proceed with the work until the proper authorizing document has been issued by the owner.

Inspections

The inspection procedures vary widely among projects. In general, the idea is that work can't be covered until inspected. Let us get into the details.

Here is the list of the variety of inspection processes out there:

Private-Sector Project under City or County Inspection

In this case, the city or county issues an inspection card that remains on site. The card shows the required inspections. The contractor calls for inspection for each item to be signed before moving to the following step. For example, inspect the bottom of a foundation trench before proceeding with placing the rebar. The property can't be occupied until the final inspection is signed off.

A Public Works Project like a Community Center

In this case, you can have a combination of inspectors, one from the city and another inspector from the public agency that owns the place. In this case, the contractor has to follow the city's and the owner's inspection procedures.

A Public School Project under the Jurisdiction of the State

For example, in California, the school designs have to be reviewed and approved by the Division of the State Architect (DSA). They review fire and life safety, structural, and accessibility. In this case, the school district has to employ a DSA-certified inspector who will be the main Inspector of Record (IOR). The IOR may have specialty inspector's assistance like roofing, plumbing, etc. In this case, there may not be any city inspections other than the IOR.

Hospital Projects under the Jurisdiction of the Office of Statewide Health Planning and Development (OSHPD)

OSHPD is the same process set for item 3 for DSA school projects.

Make sure you ask about the inspection process, the inspection requests forms, and advance notice needed at the beginning of the project. For example, some inspectors may require forty-eight hours before coming out for inspection.

So what are some of the inspections that may be required? Let's take a new construction project for example.

- The first task is grading to get ready for the underground utilities and foundation. You will probably have an inspection for the excavation bottoms and the compaction of the various lifts as you go up to the pad elevation.
- The pad has to be certified before more trenching is done.
- Inspect underground utility lines before backfilling.
- Inspect the bottom of the foundation trenches before the rebars are placed.
- Inspect the foundations rebars before pouring the foundation concrete. The same for the concrete slab on grade.
- Wall framing inspection.
- Rough wall plumbing, electrical inspection before the wall can be closed.
- Etc.

At the end of the project, before the facility can be accepted and occupied, there are typically several tests and inspections that have to be conducted. It would be smart to start paying attention to these requirements early on, so when it is time, the occupancy of the project is not delayed. Here are some examples:

Testing and Commissioning
- Commissioning is the process of testing all the installed systems to verify that they are performing appropriately; for example, HVAC units, irrigation system, etc.
- There are a lot of forms and pre-tests that would be required to complete this task. It behooves the contractor to become aware of this process early.

Air Balance
The air-conditioning system has to be air-balanced, that is, measuring the flow in the ducts to verify that it is appropriate. This will require all the finishes to be complete. That may take several days.

Owner's Training

Most specifications require the contractor to conduct training for the facility's end users. For example, how to handle the fire alarm system, how to operate the HVAC system, etc.

Real-Life Gauge

You already know how I feel about the importance of the human factor in managing projects. Well, managing the inspectors is in the heart of this. The inspector is a human with an ego, personality, attitude, and baggage from previous bad projects.

You want to do all you can to establish a good relationship with the inspector. Don't call him when you're not ready; he/she is not your superintendent or quality control person. The easier you make the inspector's life, the smoother and faster the project goes.

Get to know the inspector, benefit from his/her experience, and accept his/her advice. Make him/her feel part of the team. His/her advice and early comments will help the project a lot. Address all their comments. Make them feel that their observations are being taken seriously.

The inspector may have been a previous plumbing contractor. That means expect the most scrutiny and comments to be on plumbing issues. Plumbing inspections may be the inspector's moment to shine and show his/her strength.

I have seen projects drag because the inspector is so upset at the contractor that he/she made their life miserable on site, and every screw is an issue. On the other hand, I have seen great relationships with inspectors that will cut the contractors a lot of slack in approving work. I am not saying they are ignoring deficiencies.

It is always a good idea when you get an inspector on site to ask them if they see anything wrong in advance or ask them what you should do to make their next inspection go well. Their advice is priceless.

Best Practices

• Do not call for inspection if the work is not really ready. That will upset the inspector for wasting his/her time.

• Keep copies of all inspection requests in a chronological sequence. That will tell the story in case of a delay claim.

• Log the open correction items so you can track the issues.

Pay Requests – Retention

Money makes the world go round. Heard that before? That is certainly the lifeblood of construction projects. Hundreds of thousands get spent on projects every month, laborers get paid weekly, and suppliers have to get deposits before releasing material and more.

Keeping the money flowing keeps the work going. Pay requests or any other form of invoicing is the vehicle for contactors to get paid.

Unfortunately, there are so many issues tied into processing payment request that make this process somewhat complicated. Let's get into the details from the beginning.

A construction project has numerous trades. So the contracts have to set a method to determine how much a contractor is owed every month so that monthly invoices are processed. That can be done in a variety of ways:

• Certain milestones are set in the contract for payments. For example, $500,000 is due at completion of foundation; $450,000 is due at completion of framing; etc.

• For projects that are set to be a cost + where the contractor gets paid actual cost plus mark-up. In this case, the contractor has to put together a monthly invoice with the backup for the actual costs and submit with the monthly pay request.

• A schedule of values is established and approved at the beginning of the project. This is the most common method. Here is how it works.

  ○ At the start of the job, the contractor breaks down the cost items of the project per trade.

  ○ This breakdown can be at any desired degree of details. For example, a line item can be simply painting, or it can be broken down into "prep," "priming," and "finish coat."

  ○ At a set date for each month, the project's team goes over all the line items for which work was done and assigns a percent for that line item.

  ○ Based on the approved percents, the amount due for that month is calculated.

  ○ See the example below. One table shows the schedule of values with approved percents, and the following table shows the pay application itself.

° Notice in the schedules below the total numbers that transferred from the Schedule of Values to the Pay Application.

ITEM NO.	DESCRIPTION OF WORK	SCHEDULED VALUE	WORK COMPLETED FROM PREVIOUS APPLICATION	WORK COMPLETED THIS PERIOD	TOTAL COMPLETED AND STORED TO DATE	% COMPLETE	BALANCE TO FINISH	RETAINAGE
	Base Contract							
1	Mobilization	$ 20,000.00	$15,000.00	$5,000.00	$20,000.00	100%	$0.00	$1,000.00
	Demolition							
2	Demo Building	$ 65,000.00	$50,000.00	$15,000.00	$65,000.00	100%	$0.00	$3,250.00
3	Other Site Demolition	$ 20,000.00	$0.00	$4,000.00	$4,000.00	20%	$16,000.00	$200.00
	Concrete work							
4	Building Foundation	$ 250,000.00	$125,000.00	$125,000.00	$250,000.00	100%	$0.00	$12,500.00
5	Building Slab on Grade	$ 95,000.00	$20,000.00	$27,500.00	$47,500.00	50%	$47,500.00	$2,375.00
6	Site Concrete	$ 800,000.00		$0.00	$0.00	0%	$800,000.00	$0.00
7	New Side Walk	$ 75,000.00		$0.00	$0.00	0%	$75,000.00	$0.00
	Electric Work							
8	Trenching for conduits	$ 120,000.00		$0.00	$0.00	0%	$120,000.00	$0.00
9	Haul Off spoils	$ 50,000.00		$0.00	$0.00	0%	$50,000.00	$0.00
10	Place conduits	$ 65,000.00		$0.00	$0.00	0%	$65,000.00	$0.00
11	Backfill Conduits	$ 45,000.00		$0.00	$0.00	0%	$45,000.00	$0.00
12	Electric Vault	$ 15,000.00		$0.00	$0.00	0%	$15,000.00	$0.00
13	Pull Wires	$ 85,000.00		$0.00	$0.00	0%	$85,000.00	$0.00
14	Painting	$ 35,000.00		$0.00	$0.00	0%	$35,000.00	$0.00
15	Plumbing	$ 450,000.00		$0.00	$0.00	0%	$450,000.00	$0.00
16	Final clean up	$ 15,000.00		$0.00	$0.00	0%	$15,000.00	$0.00
	General Conditions							
17	Project Manager	$ 120,000.00		$0.00	$0.00	0%	$120,000.00	$0.00
18	Superintendent	$ 120,000.00		$0.00	$0.00	0%	$120,000.00	$0.00
	Subtotal	$ 2,445,000.00	$ 210,000.00	$ 176,500.00	$ 386,500.00	$ 0.16	$ 2,058,500.00	$ 19,325.00

Schedule of Values

APPLICATION AND CERTIFICATE FOR PAYMENT

TO OWNER:	ABC Development, Inc. 555 Main Street USA	PROJECT:	New community Building 555 Main Street USA	Application No. 3 PERIOD TO: 2/1/16 - 2/28/16
FROM CONTRACTOR: WY Construction		VIA ARCHITECT:		

CONTRACTOR'S APPLICATION FOR PAYMENT

Application is made for payment, as shown below, in connection with the Contract. Continuation Sheet, AIA Document G703 is attached.

1. ORIGINAL CONTRACT SUM................................. $ 2,445,000.00

2. Net Change by Change Orders............................. $ (25,000.00)

3. CONTRACT SUM TO DATE(Line 1 + 2)..................... $ 2,420,000.00

4. TOTAL COMPLETED & STORED TO DATE................. $ 336,750.00

5. RETAINAGE:
   a. 10% of Completed Work $33,675.00

6. TOTAL EARNED LESS RETAINAGE.......................... $ 303,075.00
   (Line 4 less Line 5 Total)

7. LESS PREVIOUS CERTIFICATES FOR PAYMENT
   (Line 6 from prior Certificate)............................. $ -

8. CURRENT PAYMENT DUE.................................... $ 303,075.00

9. BALANCE TO FINISH, INCLUDING RETAINAGE
   (Line 3 less Line 6) $ 2,116,925.00

The undersigned Contractor certifies that to the best of the Contractor's knowledge, information and belief the Work covered by this Application for Payment has been completed in accordance with the Contract Documents, that all amounts have been paid by the Contractor for the Work for which previous Certificates for Payment were issued and payments received from the Owner, and that current payment shown herein is now due.

CONTRACTOR: By:_______________________________ Date:____________

Subscribed and sworn before me this ____day of__________, 2014

Pay Application

What the owner's CM Should Look for When Reviewing the Contractor's Pay Application?

If you are a CM working for the owner, you will be tasked with reviewing and approving the payment for the contractor. Here are the most common issues you need to look for:

- The date on the application and the period it covers. Those have to be accurate, or you may have just approved a March payment in July, and later someone comes back asking for the July payment.
- The contractor is getting overpaid. So watch for the percents for each line item to reflect reality on site.
- To avoid overpayment, you should be careful at the beginning when you are approving the schedule of values. Some contractors frontload the schedule of values. What that means is that they place more money in early trades than the later trades. If that gets approved, then even if the percents are accurate, there will be a point where the amounts paid are more than the remaining work.
- Make sure that the balance left on the job is enough for the remaining work to be completed.
- There are typical tasks that usually required of contractors when submitting a pay application:
  ○ The schedule is updated.
  ○ The as-built drawings are up-to-date.
  ○ Certified payroll reports (in public works) are up to date.
  ○ All releases have been submitted (we covered this topic previously).

Retention

Typically, a certain percentage amount is withheld from each pay application, which can be 5% or more as depicted in the contract.

For example, if the contractor is billing for $100,000, the contractor gets paid $95,000. The reason for retention is to keep some money for correcting defects toward the end of the job. Typically, when the project is at substantial completion, the owner issues the contractor a punch list for the corrective items, prices each of the items, and releases part or the complete retention at the end of the job.

Billing for Material

This is one of the areas that gets the most arguments, and different contracts have different clauses regarding how to deal with this case.

This is about contractors billing for material or equipment to be used on the job. These can be materials delivered but still not installed, material still in the subcontractor's storage but not on site, or material still being fabricated and the contractor paid a deposit for it. Owners like to pay for material and equipment already used in the project. Most owners will pay for material or equipment on site. Some will pay 75 percent of it, thinking that owners may pay for it, only to find they are gone the next week. For material or equipment off site, most owners will ask for that material to be placed in a bonded warehouse with proof before they can pay for it.

So you need to make sure to read the contract before approving a pay application. My main advice is facilitate the cash flow to contractors if you want your job to go well, of course, assuming the above warnings have been checked. I have seen owners use the pay application as a tool to punch a contractor for other irrelevant reasons using creative logic to do so. You are basically shooting yourself in the foot if you slow down the progress of the work.

Closeout

We defined in a previous chapter the definition of substantial completion. You forgot already? Okay, here it is again.

Substantial completion is when a project gets to a point where the new facility is ready for its intended use and what's remaining are minor corrective items.

The phase between this point and the final completion is the closeout phase. What takes place during this phase?

• Complete the corrective work listed in the final punch list.
• Deliver the completed as-built plans. This is the plan that has red lines showing actual location of utility lines, changes to the plans, etc.
• Deliver the manuals and warranties for the various trades as specified in the contract documents.

• Complete closeout requirements from the owner or relevant government agencies having jurisdiction over the project.
• Any other requirements outlined in the contract documents for this phase.

Usually by this time, the members of the project team are tired of the project, and the subcontractors are focusing on the new project, and everyone gets annoyed by the demands of this phase.

The best way to handle this phase is to start early, when subs are still on site and momentum is still high. So here are some tips:

• Allocate one person for tracking and collecting the closeout documents.
• Review the contract documents and prepare a closeout checklist to track the required items. Usually, each specification section will have a closeout paragraph in that section.
• Start at least three months ahead of the substantial completion to ask subcontractors to provide the needed documents.
• Work with the inspector to note corrective items as you go, so the punch list comesoutshortandnot550items.Theownerhastoencouragesuchpractice.

Demobilization

As the term indicates, this means packing up and leaving the site:

• Removing the site office and other temporary facilities, like fences and temporary toilets.
• Remove any equipment from the site.
• Clean up where you messed up the place and had your site office, storage bin, and toilet.

Congratulations! Your project is done, and you are off to the next project to deal with all what we talked about so far all over again. Come on, it isn't that depressing. There is great satisfaction to see your project done, and the owner had a big opening ceremony with great speeches where they mention everybody but you.

Done? Not so fast.

There is a warranty period after the completion of the project for one year, two, or three, where the owner needs to call you to take care of warranty items. For example, an air-conditioning unit stopped working or there is lead in the meeting room, etc.

What you do here is call the right subcontractor to send people to take care of the problem.

And now you have finished!

Chapter 10
Scheduling

Every person is a scheduler. Don't you plan what you are going to do the next day or the following week? Congratulations, you are a scheduler.

I know it is a bit more complicated, but a construction schedule is about the same thing, which is planning what is the sequence of activities that you plan to follow to build the project.

There are so many excellent books that go in-depth about the theory of scheduling. I am not going to attempt to do that here. You can buy any of these books for that purpose. What you will get in this chapter is practical information about several aspects of scheduling in addition to learning basic principles.

A schedule is like a road map:

- You set the overall route (baseline schedule).
- You start driving (the construction starts).
- You hit a closed road (a problem occurs on site or a delay was caused).
- You find an alternate route to reach the same place (you develop a recovery schedule).
- Without it, you may be lost (monitor and update the schedule to stay on track).

Here is another analogy that will show you that you are a scheduler, and you don't know it yet.

If you are preparing a birthday party, what do you do?

- You make a to-do list (develop a list of scheduled activities).
- You think what to do first then second (develop the schedule logic).
- Check on the party plan's progress (prepare a schedule update).
- As you are writing the invitations, your six-year-old decides it is time for a sandwich. You stop what you are doing to take care of

him and get back to task (a delay occurred that impacted a critical activity, which delayed the whole schedule).

• Speed up one of the plan's steps to be ready on time (accelerated the schedule to maintain the target date).

We will get into the definitions of these terms later in the chapter. But did you get an idea what a schedule is all about?

Preparing a construction schedule requires the following:

- Knowledge of construction process.
- Good judgment calls.
- Planning and vision.
- Mathematical and logical mind.
- Knowledge of trades.

Let's get started.

History

In the early 1910s, Mr. Henry Gantt came up with a way to graphically represent a timeline of the planned activities by showing horizontal bars spanning across a time line (the Gantt chart). We still use that bar-chart style today as the best way to represent a schedule.

The problem with that is that there are no interrelations and dependence built in that tied activities. So if one activity is delayed and you extend that bar line, you will have to figure out the impact on the other activities manually versus an automatic update. Imagine doing that to a schedule of two thousand activities, for example. That is a very time-consuming task and so impractical that makes this missing feature a disadvantage.

How about if you would like to perform a "what if" scenario? For example, you want to find out if eliminating the west wing restroom from the scope will have any impact on the overall schedule. Adjusting all the activities manually may not be a practical task.

In 1956, DuPont de Nemours, Inc. started exploring the possibility of producing computerized schedules. Mathematicians worked on getting the computer to generate the schedule if it was fed information about the duration of each activity and sequence of work.

That was the basis of today's Critical Path Method (CPM) scheduling, which we will get into the principles of shortly.

This new CPM method was tested successfully in several large projects. The new method was adopted by the navy developing the Polaris missiles in 1958. They developed their own technique called Performance Evaluation and Review Technique (PERT). The project involved 250 prime contractors and 9,000 subcontractors. The aerospace industry used PERT after the Navy's success.

How does PERT compare to CPM?

• CPM uses definite durations, while PERT has three durations for each activity, optimistic, most likely, to most pessimistic durations.

• PERT was based on the statistical information of most likely durations.

• The problem with PERT is the three dates, optimistic, most likely, and most pessimistic durations.

• Three possible completions for each activity. The computations of the many possibilities of cost and time were too much for the computers at that time.

In using the CPM method, each activity is assigned on duration. That fact plus the CPM calculations made CPM much more practical that people started moving away from the cumbersome PERT.

Basic Terms

Before we get into the basic principles of CPM scheduling, let me introduce you to some of the basic terms that you need to know, whether you work as a scheduler or if you are simply reviewing or discussing a schedule.

• Baseline Schedule. This is the first schedule that the contractor submits to the owner. It is the initial vision and plan showing how the contractor perceives how the project will progress. Once approved, it becomes the basis of the future updates and delay issues. That is why an owner's representative has to review this submittal carefully before approving it.

• Progress Update Schedule. Typically, the contractor is required to submit at the end of each month a schedule update showing the actual progress up to that date and projected completion moving forward. If there are any delays, they have to be shown on the schedule update with a narrative explanation.

• Data Date. This is a date that is set as the basis for the schedule update.

• Recovery Schedule. If the project gets delayed for reasons caused by the contractor, the contractor has to prepare a schedule that shows how they will bring the project back to its planned completion date.

• Impacted Schedule. If the project has been delayed for valid reasons not caused by the contractor, the contractor has to insert these delay activities in the schedule to show the impact of this delay. Hence the term *impacted schedule*.

• What-If Schedule. This is examining the impact on the schedule for a hypothetical scenario. For example, what if we start the plumbing a month earlier than what is shown, or what if we add another elevator to the scope of work? What would the impact on the schedule be? That is a "what if" schedule. CPM makes this very easy to perform.

• Activity. Activities are the various work tasks listed on the schedule. For example, excavate for footings, place footings rebar, pour footings, etc. Each of those is an activity.

• Original Duration. When an activity is added to the first schedule, the baseline schedule, or if a new activity is added later in an update, duration is assigned to that activity. Since this is the first shot at assigning a duration for that activity, it is called *original duration*.

• Remaining Duration. When you are doing a schedule update on June 30, for example, and you have an activity that started in May and is still ongoing, the schedule update will show the actual start date of this activity in May and the number of days projected beyond June 30 until its completion. This is the "remaining duration" of that activity.

• Resource Loading. The resources for performing an activity, let's say excavation of footings, are labor, material, and equipment. Most scheduling software allows you to assign the number of laborers, equipment, and material to be used for that activity. This process is "resource loading" for that activity. The benefit is providing information about the resource needed, how much has been spent, and how much more is still needed. You can plot histograms of number of resources across time that is needed. The information provides valuable planning and management tools.

• Cost Loading. Similar to the above resource loading task, you can assign a cost to each activity, which will provide the same benefits but showing cost data. That provides information about how much cost was spent, how much is left to spend, and plot cash flow charts.

• Critical Activity. A critical activity is an activity that does not have any leeway to be delayed a single day. A one-day delay on a critical activity impacts the completion date of the whole project.

• Critical Path. The set of activities on the schedule where each activity is a critical activity is called a critical path. Any delay to the critical path will impact the overall completion date of the schedule.

• Noncritical Activity. As the name suggests, these are activities that are not on the critical path. That means a noncritical activity has a certain number of days that can be delayed without impacting the overall completion date of the schedule. We call that total float, which we will define below. Each activity gets its own number of float days based on the schedule relationships.

• Early Start, Early Finish. Later in this chapter, we'll get into the calculations that scheduling software perform to calculate the various dates of the activities and the overall schedule. It will also calculate which activities are critical and which ones are not. In the process of these calculations, the software will calculate the "early start" of each activity. The "early start" of an activity is the earliest date the activity can start based on durations and inserted relationships. For example, the earliest date the drywall can be started is after the insulation activity is completed because that was the relationship that was inserted tying these two activities together.

The "early finish" date is the earliest the drywall can be finished based on its "early start and the duration of the activity." In other words, this is the earliest it will be finished based on the inserted duration and relationship.

• Late Start, Late Finish. The "late start" of an activity is the latest date that activity can start without impacting the schedule. So obviously, the late start date and the duration will give you the "late finish." So looking at the early start and late start of an activity will tell you the earliest that activity can start and the latest date it can start without impacting the schedule. It's a pretty powerful management tool for a project manager. It allows you to track activities and look ahead at the upcoming activities to plan the work without impacting the schedule.

• Percentage Complete. When you are doing a schedule update where an activity already started and still has a remaining duration, that activity will be a certain percentage complete. That percentage is also reflected on the assigned cost and resources. It is another planning tool.

• Total Float. The total float of an activity is the number of days that activity can be delayed before it impacts the overall schedule.

• Free Float. The amount of days an activity has before it impacts the immediate successor activity, as opposed to the total float, which is the number of days before impacting the overall schedule.

• Positive/Negative Total Float. A positive total float of five days for an activity means that the activity still has five days of slippage before it delays the overall schedule. A negative total float of minus five means that the activity is five days late.

• Predecessor/Successor. When you are building a schedule, you are listing activities in a certain sequence. For example, we complete drywall, then start painting. In this case, the drywall is the predecessor of painting. Painting is the successor of drywall. Assigning realistic durations, predecessors, and successors is key to having a realistic schedule that can be used to manage the progress of a project.

• Logical Relationship, FS, SS, FF. Take the example above about drywall and painting being tied together as predecessor/successor. That is done by entering a relationship between these activities when you are building the schedule.

1. FS: Finish–Start relationship. That means the successor activity starts after the predecessor finishes.

2. SS: Start–Start relationship. That means the successor and the predecessor will start at the same time.

3. FF: Finish–Finish relationship. That means the successor and the predecessor will finish at the same time.

• Lag Time. Let's see any of the relationships defined above. If we want the successor activity to start five days after the predecessor activity finishes, the relationship will be FS – 5. The five days is called lag time.

• Calendars. When you are building a schedule, you have to assign a certain calendar that is used to calculate the dates. For example, if the scheduling specifications calls for the definition of "DAYS" to be calendar days, that means that the duration is calculated based on seven days per week. If the specification defines it as working days, that means holidays and weekends are not counted. So an eight-day duration in calendar days means you start on Monday, for example, and the eighth day is Monday of the following week. If it is eight days in working days calendar, then the activity starts on Monday, but the eighth day will be Wednesday of the following week. The scheduling software typically has built-in calendars that you can use or customize per job.

    • Milestones. A milestone activity is an activity with zero duration. It is like a point in time. The reason is to track specific deadlines. For example, we can insert an activity called "Completion of Building C" and assign a zero duration. The reason is that it is important to track what date will Building C be done. That milestone activity makes it easy to locate on the schedule and read quickly the current projected completion date of Building C. Other examples can be "Substantial Completion," "Final Completion," etc.

    • Constraints. You can insert a constraint on an activity such as "Must Finish by 5/4/16." What that will do is tell the software to calculate the float for this activity based on a desired 5/4/16 date to track that important requirement. Other constraint examples can be "Must Start On or Before" and others that are built into the software.

How To Build a Schedule

Okay, now that you know these basic terms, how do we build a schedule? After you get yourself familiar with the scheduling software of your choice, here are the simple basics to build a schedule:

- A set of activities are identified (see example table below).
- Their durations are identified (see example table below).
- A calendar is identified. We talked about this above. Do the specs call for calendar days or working days? For the sake of this exercise, we will use calendar days.
- Now, we take the above list and build the schedule's network. There are many methods of doing this. You can refer to a scheduling book for this information. But for our purposes here, I will use my favorite method, which is the precedence diagram method (PDM).

Activity Description	Duration (Days)
Site Demolition	7
Grading	15
Foundation	20
Framing	30
Rough Electric	40
Rough Plumbing	25
Rough HVAC	20
Insulation	7
Drywall	15
Finishes	50
Exterior Irrigation	10
Exterior Landscape	7
Final Cleaning and Demobilization	2

It is simply drawing up a set of boxes, with each box representing one of the activities above. Draw them up in the sequence that you envision building the above project. Then identify the relationships between these boxes. For simplicity, we'll use the FS (Finish-Start) relationship in this example.

There is no right and wrong as long as the sequence makes sense and it corresponds to how the project will be built. The other important aspect is to have realistic durations, not just numbers that will make it look good.

Here is the network for the above activities as I see it. You may want to list them in a different sequence.

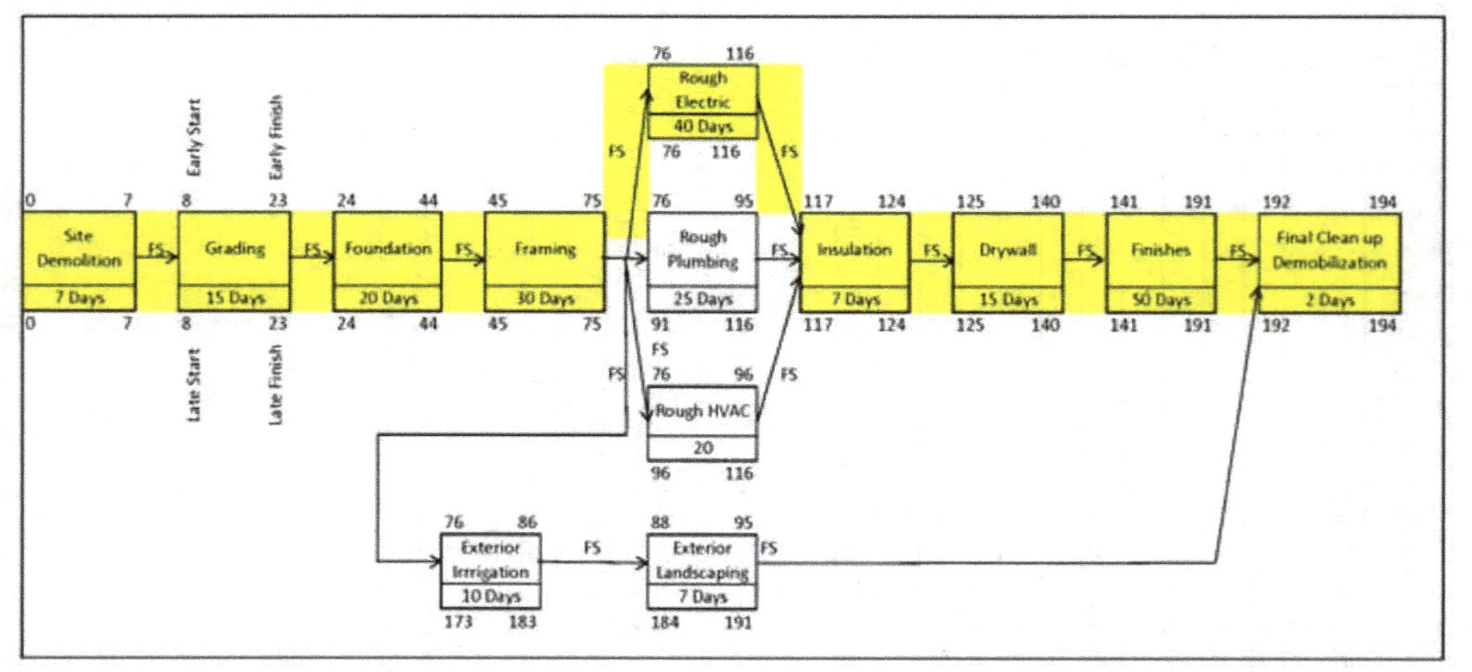
Early Start
Early Finish
Late Start
Late Finish

0 Site Demolition 7
7 Days
0 7
FS

8 Grading 23
15 Days
8 23
FS

24 Foundation 44
20 Days
24 44
FS

45 Framing 75
30 Days
45 75

76 Rough Electric 116
40 Days
76 116
FS

76 Rough Plumbing 95
25 Days
91 116
FS

76 Rough HVAC 96
20
96 116
FS

117 Insulation 124
7 Days
117 124
FS

125 Drywall 140
15 Days
125 140
FS

141 Finishes 191
50 Days
141 191
FS

192 Final Clean up Demobilization 194
2 Days
192 194

76 Exterior Irrigation 86
10 Days
173 183
FS

88 Exterior Landscaping 95
7 Days
184 191
FS

- Each box is an activity from the above list.
- Each box has the name of the activity and the duration below. For example, the second box is Grading and has a duration of fifteen days.
- The arrows show the sequence as planned by the scheduler.
- For example, Foundation will start after Grading is done.
- After Framing is done, three activities will start: Rough Electric, Rough Plumbing, and Rough HVAC.
- Insulation can't start until all three activities before it are done (that is Rough Electric, Rough Plumbing, and Rough HVAC).
- Grading is the "predecessor" of Foundation.
- Foundation is the "successor" of Grading.
- What is the successor of Framing? Look up and try to answer first. Did you get it?
- The answer is Framing has four successors: Rough Electric, Rough Plumbing, Rough HVAC, and Exterior Irrigation. That can't start until Framing is done. Why? Because I assigned a Finish-Start relationship between Framing and the four successor activities.
- The network shows that Exterior Irrigation and Exterior Landscape will happen concurrent with the top set of activities.
- This project will take 194 days to complete.

Now, what are all these other numbers on the top and bottom of these boxes?

- The top left numbers are the early starts of each activity.
- The top right numbers are the early finish of each activity.
- The bottom left numbers are the late start of each activity.
- The bottom right numbers are the late finish of each activity.

What in Heaven's name does that mean?

- Site Demolition starts on day 0.
- Its duration is 7 days.
- So the earliest time it can finish is $0 + 7 = 7$.
- Since the relationship between Site Demolition and Grading is FS (Finish-Start), the earliest time Grading can start is when Site Demolition is finished.
- So the Early Start of Grading is day 8 after Grading finishes on day 7.

- Since the earliest day Grading can start is day 8 and its duration is 15 days, its early finish would be day 23 (8 + 15 = 23).
- The rest of the early starts and early finishes are calculated the same way.
- Now let's look at Insulation.
  - The earliest is can start is when all three predecessors are completed.
  - The early finish of Rough Electric is 116.
  - The early finish of Rough Plumbing is 95.
  - The early finish of Rough HVAC is 96
  - So what is the earliest day can Insulation start?
  - That would be day 117, after Rough Electric is done since it is the longest.

Read this again. I know it is confusing.

- This is called Forward Pass Calculation.
- The lower numbers are the Late Start and Finish dates. The above calculation gave us the earliest dates that activities can start and finish. The Late Start and Late Finish dates give us the latest dates that these activities can start and finish without impacting the end date at 194 days.
- This is done by starting calculation from the end moving backward, which is called the Backward Pass Calculation.
- It is simply saying we're going to fix the finish date of the last activity at 194, which we can't exceed, and then calculate backward the latest date of the other activities to meet this objective.
- So the Late Finish Date of the activity "Final Clean Up and Demobilization" is at 194 days. 194 − the duration of 2 = 192, which would be the latest day it can start since the latest finish is day 194.
- Well, if the latest day is Final Cleanup and start, then the latest day the "Finishes" activity can finish is day 191, then 191 − a duration of 50 days = 141, which is when the latest day "Finishes" can start.
- Now let's take the three activities: Rough Electric, Rough Plumbing, and Rough HVAC. What would be the latest day these can finish if Insulation's latest start date is day 117?
- Because of the FS (Finish-Start) relationship we set between those three and Insulation, the latest day all three can finish would be day 116.

Okay, so what does all that gibberish mean?

- Take a look at the top and bottom numbers of the activity "Rough HVAC." The earliest day this can start is day 75, and the latest day it can start without impacting the schedule is day 96. That means if this activity gets delayed from starting on day 75 and starts no later than day 96, the overall schedule will not be impacted.
- That means it has 96 – 75 = 21 days it can slide. This is what we call Total Float.
- So Total Float = LS – LF or LF – EF.
- We also call the activity "Rough HVAC" a Noncritical activity.
- Look at the activity "Framing." The late and early dates are exactly the same. That means this activity does not have any room to be delayed. That makes this activity a Critical Activity.
- The above chart highlighted all the activities that are critical. If you follow the line of highlighted activities from left to right, that is one of the paths from start to finish. Since this is the path that goes through all critical activities, we call that a Critical Path, which is the longest path on the project.
- So when you are managing the project, you need to make sure that none of the Critical activities along the Critical Path are delayed to avoid an overall delay to the project.

Since the progress on site often happens differently than the original schedule, the critical path may change from time to time, depending on the duration of the various activities. For example, we said above that Rough HVAC has a Total Float of 21 days. If Rough HVAC is delayed 40 days, then the new schedule calculation may result in showing the new Rough HVAC activity is a Critical Activity.

On a large schedule, you may have more than one critical path. As the project manager, you need to monitor the critical activities to avoid any delays.

You also need to look at the Near-Critical activities. These are activities with little float, like two days for example. That means these may not critical now, but are very close to becoming critical if they get delayed for more than two days.

Are you getting the picture of how CPM (Critical Path Method) scheduling is important for managing projects?

At this point, you may be asking yourself, "Are you telling me

I have to do the above calculations if I have a project with three thousand activities?"

No.

The scheduling software will do it for you. All you have to do is build your diagram, insert the activities, the durations, and relationships, click one button, and *boom*, you got yourself a CPM schedule.

What happens though is you get a schedule that often finishes beyond the target date. So you have to go back and change some durations and relationships to get it back on track. For example, instead of having an FS relationship between two activities, you may want to change that to SS-10, which starts the second activity ten days after the first one starts, and you don't have to wait until it finishes. That will push the final date earlier.

Let's translate all that into action. I'm going to take the network diagram above and insert it in Primavera, one of the most common software in the market, and let's see how the above schedule will look in a typical bar chart format.

Okay, I've done it. Guess how long it took me to do the following schedule? Eight minutes.

06-Apr-16

Activity ID	Original Activity Name	Start	Finish	Total Float
A1000	7 Site Demolition	05-Apr-16	11-Apr-16	0
A1010	15 Grading	12-Apr-16	26-Apr-16	0
A1020	20 Foundation	27-Apr-16	16-May-16	0
A1030	30 Framing	17-May-16	15-Jun-16	0
A1110	10 Exterior Irrigation	17-May-16	26-May-16	125
A1120	7 Exterior Landscaping	27-May-16	02-Jun-16	125
A1040	40 Rough Electric	16-Jun-16	25-Jul-16	0
A1050	25 Rough Plumbing	16-Jun-16	10-Jul-16	15
A1060	20 Rough HVAC	16-Jun-16	05-Jul-16	20
A1070	7 Insulation	26-Jul-16	01-Aug-16	0
A1080	15 Drywall	02-Aug-16	16-Aug-16	0
A1090	50 Finishes	17-Aug-16	05-Oct-16	0
A1100	2 Final Clean Up / Demobilization	06-Oct-16	07-Oct-16	0

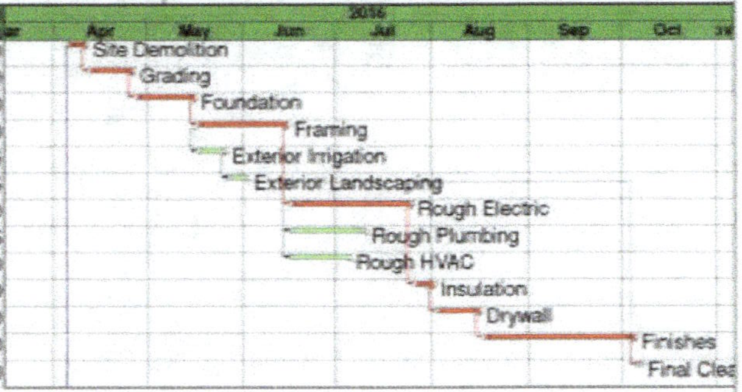

Actual Work   ◆ Milestone
Remaining Work
Critical Remaining Work

Sample Schedule for Above Network

TASK filter: All Activities

Look at what this schedule shows:

- The activities' IDs.
- The activities from the network.
- The Early Start and Finish dates.
- The Total Float of each activity.
- The critical activities are shown in red. These have 0 total float.
- The noncritical activities are green. Each one has a different total float. For example, Rough HVAC has twenty float days. That means this activity can slide twenty days without impacting the final completion date.
- If you follow the red activities, you will see the critical path. Compare that to the critical path we calculated above. Compare to the highlighted path in our network. Are these the same activities?

Narrative Report

When a schedule is submitted, typical specifications request a narrative report to go along with it. The intent is to explain the overall schedule to the reviewer and mention specific aspects of it, like a delay or sequence of events, phasing, etc.

The narrative report usually includes the following information:

- Overall background about the project
- Phasing
- Activity Codes
- Milestones
- How the schedule is organized.
- General logic
- Critical activities
- Narrative explanation about specific set of tasks (for example, the logic explaining how the closeout phase is planned).
- Schedule updates, delay analysis, and any other schedule may need a narrative report to explain.
- If this is a narrative for a recovery schedule, for example, it may include the following:
    - Same overview of the project.
    - New activities added.
    - Changes in logic.
    - Impacting activities, if any.

• If there is a delay, explain what happened and how it impacted the work.

Schedule Update

Most specifications require that the contractor submits a schedule update once a month.

The scheduler simply takes that last month's approved schedule and inserts the current status of all the activities as of the date of the update, which is called "Data Date." The update will show actual start and finish dates and projected dates of activities.

I am going to perform an update on the above schedule with a data date of May 31, 2016. In this update, the framing was delayed. Look at what the update will look like.

Activity ID	Original	Activity Name	Start	Finish	Total Float
A1000	7	Site Demolition	05-Apr-16 A	11-Apr-16 A	
A1010	15	Grading	12-Apr-16 A	26-Apr-16 A	
A1020	20	Foundation	27-Apr-16 A	16-May-16 A	
A1030	30	Framing	17-May-16 A	23-Jun-16	-9
A1110	10	Exterior Irrigation	31-May-16	09-Jun-16	110
A1120	7	Exterior Landscaping	10-Jun-16	16-Jun-16	110
A1040	40	Rough Electric	24-Jun-16	02-Aug-16	-9
A1050	25	Rough Plumbing	24-Jun-16	18-Jul-16	6
A1060	20	Rough HVAC	24-Jun-16	13-Jul-16	11
A1070	7	Insulation	03-Aug-16	09-Aug-16	-9
A1080	15	Drywall	10-Aug-16	24-Aug-16	-9
A1090	50	Finishes	25-Aug-16	13-Oct-16	-9
A1100	2	Final Clean Up / Demobilization	14-Oct-16	15-Oct-16	-9

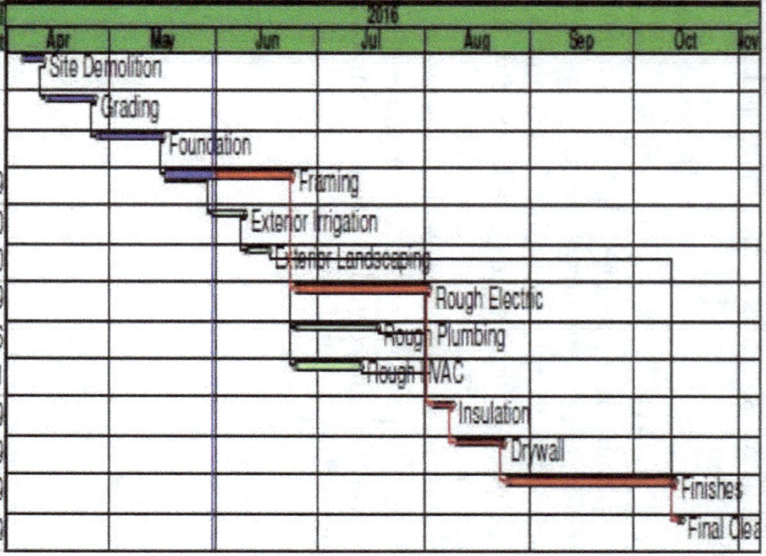

Please take a note of the following:

• The blue vertical line is at the end of May, which shows you the Data Date of this update.
• The blue activity bars to the left of the Data Date indicate the actual progress of activities up to that date. Notice how the schedule shows the letter "A" in the date columns. The "A" means "Actual." So Grading actually started on 4/12/16 and actually finished on 4/26/16. These dates are the "Actual Start" and "Actual Finish" dates of the Grading activity.
• Notice that the Framing activity started on 5/17/16, but the expected finish is 6/23/16. If you compare the schedule update to the original schedule few pages before, you will see that the original planned Finish Date of framing was 6/15/16.
• Since Framing is a Critical Activity, then any delay in Framing will impact the whole schedule.
• The Original Completion date of the schedule was 10/7/16. Now, the update shows a projected completion of 10/15/16. The project is delayed. That is why the Float Column shows (-9). Remember, we talked about that earlier. A negative float means that the schedule is late by that much.

So now, as a project manager, what do you need to do to bring this back to the original completion date of 10/7/16? Think about it and email your answer to arcline@earthlink.net.

This is why an accurate update is important, so you can track and manage the timeline of a project.

In summary, here are some tips and points to remember about a schedule update:

• Schedule update is a very significant and valuable tool for managing projects.
• Do it right because once the update is approved by the owner, it acts as the new baseline moving forward, which can be critical to any delay discussion.
• Unfortunately a lot of contractors look at schedule as a forced expense and a task that they have to do. So they do not spend time on checking the information that goes into it.
• If you don't do a good update, how will you measure the status of the project and decide what to do to keep it on track?

• If the produced schedule update is not close to reality, the schedule will be seen as unrealistic and useless. The result: Work will go on without a guiding document.

• The basic data for an update is Actual start and/or Finish, percentage complete, remaining duration, logic change, and new activities.

• Actual dates have to be backed up by documented facts like daily reports. Future projections may be based on a judgment call, like the remaining duration.

• Before you do an update, don't forget to copy and paste the last schedule. You do not want to ruin the previous record.

• What is Claim Digger? Claim Digger is a software that compares the differences between two schedules.

• The most important information out of an update is what activities form the critical path at this time. Are the milestones still on track, or has the schedule slipped?

• If you are reviewing a schedule as an owner's representative, you need to watch out for issues that impacted the progress like the following:
  ○ Look at the milestone dates to note delays.
  ○ Look at the changes in logic and added activities.
  ○ Look at the Critical and Near Critical Path Activities.
  ○ Read the update narrative report.
  ○ Does the information reflect reality?
  ○ Check the Claim Digger report.

• Keep the acceptance and rejection process practical. Use common sense; don't make it burdensome to keep things moving forward.

• If you reject the submitted schedule, state why, and the reasons have to be contractually sound.

If you are reviewing a schedule for approval, what do you need to look for? Once you approve the submitted schedule, it becomes an important contract document that can be used as the basis for future delay claims. So you need to know what to look for before you approve it. Here are some key elements to watch out for:

• Are the start and finish dates consistent with the contract?
• Are the milestone dates per contract?
• Is the phasing per contract?

- Does it include all specified requirements?
- Are all activities included?
- Is the sequence logical?
- Are the critical activities expected to be critical?
- Are all activities shown critical? (Big *no*)
- Are the durations reasonable?
- Are the trades divided into short-duration activities to track the various tasks? For example, an activity (Plumbing) that is only shown as one-time bar of 250 days is not effective to track. That has to be broken down into the various tasks of the plumbing scope. For example, excavation, place sewer pipes, backfilling, interior rough plumbing, etc.
- Minor detail, but does the title sheet read the correct project's name?
- Be practical. After all, a schedule is an initial plan.

Best Practices

Before giving you best practices ideas, I would like to stress this rule that I live by: *A schedule that is not visible is useless.*

On my job sites, you will always see a schedule posted on the wall so the deadlines and projected activities are visible for the team to follow and keep in sight. It is also a regular discussion on site to see where we stand, what can be done to beat the schedule, and forecast potential issues.

If the schedule is in a file cabinet somewhere, then the site will be running without a time guide, something like driving to a new location without a road map.

Here are some best practices pointers:

- In planning a schedule, specify reasonable durations and realistic logic relationships. Don't just make it look good on paper when reality is going to be something else. If you do that, then the schedule will become simply a bureaucratic document for compliance purposes only.
- Get the input of involved parties, like subcontractors and the superintendent.
- It is important for the superintendent to buy into the schedule so he/she feels a sense of ownership of the goal of meeting those dates. Otherwise, the superintendent's attitude is going to be

dismissive of the schedule, and it won't work as a management tool.

• Realize the fact that a schedule is a living document, and dates will change. However, it is important to keep looking at the target dates and other options in case some task gets delayed.

• Develop a system to record actual dates and impacting events to keep the updates realistic and accurate. One good way is to hang the schedule on the site office's wall and red mark actual dates as they occur. Do not wait for the update date and then go down memory lane to guess when framing actually started.

• Keep the schedule visible on site so the team can discuss where the project is and what is projected forward.

• In case of delays, document them as they occur (contemporaneously).

• If the scheduler is not on the job or he/she is a consultant, develop a good system of communicating updates and information. It is not a bad idea if that scheduler visits the site to get familiar with the work.

Real-Life Gauge

Best definition of a baseline schedule that I read: "It is the intended plan to execute the activities for the most likely durations following this most-probable sequence unless something comes up."

I am a big advocate for using schedules as a main tool for managing the project. Unfortunately, I see a lot of bad practices and approaches to schedules that turn this wonderful tool into a bureaucratic process often used as tools for disputes and fights. I have seen projects six months into construction, and the baseline schedule isn't approved yet. The construction is going on, and concurrently, the owner's and the contractor's schedulers are meeting, dissecting the guts out of the schedule, arguing in a bureaucratic bubble that has nothing to do with reality. I can understand the thinking behind this since the schedule will become the main reference to potential delay disputes, and the owner needs to protect himself. But, come on, give me a break. Maintain common sense, and move things forward.

Okay, now I feel like ranting a bit. I have told you above about the proper practice for scheduling. Here are some examples of bad scheduling practices that I have seen, which take away the value of this magnificent tool.

Example No. 1

A large owner—like a city, school district, or others—has a large construction program, and they are putting together the execution plan and preparing specifications that they intend to use in the bid documents.

One of them is the scheduling specification. The owner hires a scheduling consultant to help develop the specifications and methods to monitor the schedule during construction. So they hire this guy who knows the scheduling software inside out. Let's hope he/she ran a project or two before. This consultant now has a client to impress, so he/she pours his/her brains out on paper, thinking of every requirement and loophole imaginable and do his/her best to produce a specification to defend their client.

What we'll get is a seventy-page schedule submittal that needs another equally sophisticated scheduler to be working with the future contractors to decipher this document.

This specification section finds its way to every project's bid document that will be advertised in this program, whether the project is a seven-month $1 million project or a two-year $30 millionproject. And it has all the bells and whistles and few threats, like holding the pay application of the contractor if all these requirements aren't complied with. Because, as you know, nothing keeps a construction project going better than holding cash flow, right?!

The reality is that smaller contractors don't have full-time schedulers on board that they can afford to keep full time. Looking at this specification section written in hieroglyphic language, they decide to hire a consultant scheduler who can speak the same language. This guy who does scheduling for a living can be a hundred miles away.

Congratulations, the disconnect between the project and the schedule was just born. That remote scheduler doesn't know the project, so the contractor's project manager's time gets taken away just to feed the remote scheduler information while preparing to mobilize and start the actual work on site.

That contractor gets frustrated and tells the scheduler, "Just prepare whatever they need to get this submitted and approved."

Meanwhile, the project starts.

The submittal package comes in. It looks beautiful, professional, and is so thick that it can be a textbook by itself. It will have charts, tables, narratives, and more. The contractor simply forwards it to the owner. The owner's scheduling consultant gets tears in his/her eyes at the beautiful sight of these schedules and charts. The owner's consultant looks into every relationship and constraint that was built in and produces another large response of what is wrong with the submitted schedule. The contactor gets that report and goes, "What the heck does all this mean? I'll just send it to my scheduler and let him/her handle it." These two schedulers will communicate back and forth until the baseline schedule is approved.

And this is how projects go for six months before the schedule gets approved. Folks, you have to keep processes practical and reasonable. All schedules will change as the project moves forward.

If the above story doesn't convince you, see the table below where I took two scheduling specifications, one complicated and one simple. I outlined the required tasks of both and put a cost to each step. Compare the costs, and see how much money can be saved by being practical.

Task	Hrs	Rate	Cost
Submit a Preliminary schedule for the first 90 days	40	$100.00	$4,000.00
Submit the complete baseline schedule	20	$100.00	$2,000.00
Update the prelim schedule first 3 months	48	$100.00	$4,800.00
Prepare narrative and other baseline reports	16	$100.00	$1,600.00
Meeting to discuss the baseline	8	$100.00	$800.00
Monthly updates (12)	120	$100.00	$12,000.00
Meeting with Owner to discuss each update (12)	96	$100.00	$9,600.00
Attend weekly meetings	104	$100.00	$10,400.00
At each weekly meeting submit a 4 weeks look ahead	104	$100.00	$10,400.00
Recovery schedules	176	$100.00	$17,600.00
Fragnets in case of time extensions	20	$100.00	$2,000.00
Total			$75,200.00
Smaller Specification Section			
Prepare a baseline schedule	30	$100.00	$3,000.00
Baseline reports	10	$100.00	$1,000.00
Prepare monthly updates (12)	48	$100.00	$4,800.00
Prepare 3 weeks lookaheads weekly	104	$100.00	$10,400.00
Recovery schedule	20	$100.00	$2,000.00
Fragnets	20	$100.00	$2,000.00
Total			$23,200.00
Savings per project			$52,000.00
If you have a 500 projects program, you save			$26,000,000.00
If you have a 1000 projects program, you save			$52,000,000.00

Example No. 2

I also have a problem with contractors who don't appreciate the benefit of sound scheduling practices and treat it as a submittal that no one looks at on site. Then they wonder why the project is late and why they have no backup to justify their time extension request.

I know, I know. You may be saying, "I've been building for thirty years, and I don't need no stinking schedule to know what to do."

That may be fine, and it may work for some projects. But if you are doing a public works project, or a large project with 3,500 activities, or a problematic project with many delays, you can lose track of the overall timeline, and you may jeopardize your delay claim, which lacks a proper schedule analysis as required in the specifications.

So the contractor submits a schedule that gets approved and shoves it in the file cabinet. When it is time to do the schedule update, they take the last approved schedule out of the file cabinet, clean the mustard stains, and try to remember the dates to use for the update. They insert any dates, just to get the update submittal going, thus accumulating the problem and making the schedule even more useless.

Please review the best practices above and apply sound practices. It will help you a lot.

Chapter 11
Delay Analysis

Construction projects often suffer from delays due to a wide variety of reasons, which can have severe financial impacts on the project. As a result, delay claims may be filed. The analysis of the delay impact and the causes and effects of the delaying activities is one of the most complicated types of claims analysis. It requires an expert with extensive knowledge of construction projects, means and methods, scheduling, and the ability to develop a sound methodology to conduct the analysis. I have had many delay claims to analyze and negotiate years after the projects were done. This resulted in a detailed and intensive research of the documents to verify schedules, events, sequence of work, changes during construction, and the delay impact. In some cases, I had to rebuild schedules to see what went on.

This chapter will address these challenges and the various delay analysis methods. Below is a list of common delay causes encountered on construction projects. One of the complications of a delay analysis is that the delays can be caused by few of these listed causes or a complex mix of these causes. The time of their occurrence and who caused what delay add to the difficulty of the analysis.

- Errors and omissions in the contract documents
    - Missing information
    - Wrong details
    - Not having a phasing plan in the bid documents when the site work has to be done in phases.
    - Conflicting information that need design revisions.
- Contractor-caused delays for reasons under their control
    - Not having enough labor force on site.
    - Contractual problems between the prime contractor and subcontractors.
    - Cash flow issues.
    - Lack of proper planning and management of the project.

- Delays for reasons beyond the contractor or owner's control
  - Strikes
  - Out-of-state manufacturer's shutdown.
  - A subcontractor going out of business in the middle of the project.
  - Unusual weather conditions.
- Owner-caused delays for reasons under their control
  - Scope changes
  - Limiting contractor's access to parts of the site.
  - Cash flow
  - Late processing of contractor's requests for clarifications and change orders.
  - A higher-level political factor that impacted the project's progress.
- Personality conflicts between the project's team
  - Unfortunately, sometimes this factor results in the team making things difficult on site that cause delays. In this case, each party blames the other for the delay.

One of the main steps in the delay analysis is to research the project's documents to identify causes like the above that delayed the project. The methodology used to determine the impact of these factors is the heart of the difficulty of this type of analysis. To better understand the level of difficulty involved, please note the following basic concepts that have to be factored in:

Critical, Noncritical Delays, and Float

The project activities in a schedule are two types: critical and noncritical. The noncritical activities have certain number of days (float) where the activity can be delayed without delaying the whole project. For example, five days float means that the activity can be delayed up to five days without delaying the whole project. The critical activities have zero or less float, which means that each delay day will delay the whole project.

Determining which activities are critical and noncritical depends on the durations and logic of the sequence of activities. Rebuilding the schedule after the fact, determining which activities are critical and which ones are noncritical, and establishing the logic, which usually changes through the project, takes a highly-technical research of the documents. Some assumptions and judgments may have to be taken during the analysis.

Excusable and Nonexcusable Delays

Excusable delays simply mean delays at no fault to the contractor. In this case, a time extension is owed to the contractor. Nonexcusable delays are delays due to the contractor's fault. A detailed revision of the contract's terms and conditions is critical to properly classify the type of each delay identified in the analysis.

Compensable and Noncompensable Delays

Compensable delays are delays where the delayed party is owed money to compensate for the loss due to the delay. Noncompensable delay is a delay where a time extension is owed, but no compensation is owed to the delayed party. For example, some contracts specify that delays due to reasons beyond the control of the owner and the contractor are delays where a time extension is granted, but no compensation is paid to the delayed party. A good understanding of the contract terms is critical to the expert analyzing the delay claim.

Concurrent Delays

Some analysts simply list the delays, calculate the number of days for each delay, add them up, and claim the total as the total number of delay days. Well, that is far from being an accurate analysis. The timing of each of these delays is important. We may have three delay causes that occurred during overlapping time periods or within the same period. The schedule and the actual site events have to be examined at the start date of each one of these delays to analyze its impact. We may find that only one of the three concurrent delays had an impact on the critical path of the project. After plugging that in the updated schedule, we can find out the new completion date of the whole schedule.

Using Critical Path Method (CPM) scheduling provides analysts the needed tools to conduct a proper analysis and to understand the tremendous advantage of having CPM technology. We discussed CPM scheduling in the previous chapter.

The following are the basic different methods that are commonly used to analyze delays:

- As-planned vs. as-built method
- Impacted as-planned method

- Collapsed as-built or "but for" method
- Window analysis method
- As-built method
- Contemporaneous method

As-Planned vs. As-Built Method

The analyst compares the dates and durations of selected activities shown on the as-planned schedule with the actual dates and durations on an as-built schedule and considers the difference to be the delay on the job.

This is a very simplistic view of the delay claim because it ignores the following important factors:

- The cause of the delays.
- The timing of the individual delays and their impact on the schedule to be able to attribute the correct amount of delay days to the right responsible party.
- It ignores the impact of concurrent delays.
- It ignores the fact that the logic and sequence of the as-planned schedule may have changed through the project due to numerous delaying factors.

Impacted As-Planned Method

In this method, the analyst lists the excusable delays (or delays where time extension is owed to the contractor) and inserts the extended duration to the relevant activities. The analyst reads the revised completion date and calculates the days between this date and the as-planned completion date and determines that these are the number of days owed to the contractor. The sources of error in this method are the following:

- It ignores the actual as-built schedule and events on site.
- It assumes that the logic of the as-planned schedule reflect the reality on site.
- It ignores the inexcusable delays that may have been concurrent to some of these inserted delays, which impacts the number of days owed to the contractor.
- Since the analyst is only using the as-planned schedule, this method doesn't incorporate changes in logic and out-of-sequence work.

Collapsed As-Built or "But For" Method

In this method, the analyst takes the actual as-built schedule and takes out the duration of all the excusable delays (delays rightfully owed to the contractor). This revision forms the collapsed as-built schedule. The analyst reads the completion date on the collapsed as-built schedule and considers this date to be the completion date of the project, had the contractor not been delayed. The analyst calculates the days between the collapsed as-built and the completion date from the as-built schedule and considers these days to be the days owed to the contractor. The sources of error in this method are the following:

- It depends on the as-built schedule to be accurate.
- The excusable delays removed from the as-built schedule are assumed to be excusable without a complete analysis of these delays, the causes, and concurrencies. That means subjective assumptions and judgments have been taken and need to be examined.
- It doesn't factor in how the sequence of operation changed, any acceleration that took place, any recovery that took place because the as-built schedule is a representation of what really happened on site without addressing causes and effects of delays along the way.
- In some cases, where an as-built schedule does not exist, the analyst recreates the as-built schedule based on his/her research. This product does not reflect the planned logic of activities or the planned critical path.

Window Analysis Method

This method is based on analyzing the delay over the entire schedule, dividing it to windows with a selected duration. The most commonly used is monthly. The analyst looks at the activities within the selected window and updates the activities incorporating the delays within the selected window. Updating the selected window changes the as-planned schedule to an as-built schedule up to the end date of the selected window and becomes the basis for projecting the remaining activities from the end of the window to the completion of the project. The sources of error in this method are the following:

- Need to have accurate as-built information on the start and finish dates of the windows.

• The original base schedule has to be accurate.

• There may be delaying activities outside the selected window that have an impact.

As-Built Method

This method is used in the absence of reliable schedules on the job. In this case, the analyst recreates a schedule based on actual information. The analyst determines the logical ties between the activities to form a retrospective schedule, which becomes the basis for analyzing the effect of the delays. Durations are given to the activities based on reasonable time to finish the various activities. The delays are then inserted in the newly-created schedule and then compared with the actual as-built durations to calculate the number of delay days. Here are the sources of error in this method:

• The analyst has to be very experienced in construction means and methods.

• There is a lot of judgment calls by the analyst that need to be examined.

Contemporaneous Method

This is usually the preferred method of analyzing delays. In this method, the analyst takes a look at the schedule and actual site progress on the starting date of each delay, and then inserts the delays in the schedule. The new completion date is compared to the original completion date to determine the delay days. This way the impact of concurrent delays is incorporated; the new critical path reflects reality on site and effect of the delaying causes. Here are sources of error in this method:

• Having good documentation to reflect the actual site progress.

• Accurate schedule updates.

The analyst has to select the method to use. Each method has its advantages and problems. Sometimes the nature of the case, available time, document availability, or budget consideration influences the method selection.

Time Impact Analysis

Now, let me break down the steps of what you need to do to perform your delay analysis. Like I said before, the availability and the accuracy of the available documents is a factor. So let me explain the analysis steps in various scenarios.

Scenario 1: You are the contractor, and your documentation and scheduling updates are up-to-date and done properly.

• Something just happened that will have an impact on Framing, for example, which happens to be a critical path activity.

• First thing you do is send the owner a notice of start of delay. This notice informs the owner that an event took place that will delay the Framing and that there will be time and cost impact to the project.

• Giving notices on time is extremely important, especially if the dispute escalates into litigation. The idea of a notice is to make the other party aware and to give them a fair opportunity to fix the problem and minimize the impact.

• Immediately, start a folder with all relevant documents, costs, invoices, etc. That will become your backup for the delay request.

• Let's say the delay cause happened on April 10, 2016. Here is how you find out the impact:

    ° Update the schedule up to April 9, 2016, or use the latest approved schedule update at the end of March 2016.

    ° Check the completion date of the project based on this update. Let's say it is October 1, 2017, and the schedule shows a negative float of -20 days, which says that currently the project is behind twenty days.

    ° Insert the delay activity and tie it to the impacted Framing activity.

    ° Run the schedule again. Let's say the new completion day is October 20, 2017, and the total float changes to -39.

    ° That documents that the impact of this delay is 19 days (39-20).

Scenario 2: The contractor did not do their paperwork properly, and no schedule updates were done or only a few updates were done.

In this case, you'll need a more sophisticated analysis where you may have to rebuild schedules based on actual progress or go back to the latest approved schedule and insert the delay events to see the impact.

This obviously opens the door for more arguments and disputes, which may escalate into litigation.

Now, if you are the owner's consultant/construction manager, you will have to understand the above principles in order to respond to the time delay claim and check if proper scheduling practices have been applied. A lot of owners hire delay experts to analyze the submitted claim.

Concurrent Delay

This issue comes up in most delay arguments. Consider the above issue of having a delay by the owner that impacted the Framing, a critical path activity. Let's say that the contractor at the same time of the framing delay caused a delay to another critical path activity. So now you have a delay by the owner and a delay by the contractor that happened at the same time, and both impacted the critical path. This is a "concurrent delay."

The way to determine the delay entitlements is governed by each of the contract documents, which can be different.

One common outcome of a critical delay case is that the contractor gets the days but no compensation (Uncompensable – Excused Delay).

Other contracts may state that no time extension is granted in cases of concurrent delays. So you can imagine the complexity of the arguments when each party is pointing out to the other party what they did wrong.

That is why documenting delays and updating schedules as delay causes occur is the best way of dealing with delays. This is what is referred above as the Contemporaneous Method.

Best Practices

Needless to say that the best thing is not to get into delays, but unfortunately, this is a part of the reality of most construction projects. Here are some best practices tips:

- Try to foresee problems and pre-plan for work to avoid delays.
- Be proactive in teaming with the rest of the project's parties to

solve problems as they occur.

• *Establish a nice working relationship* on site. That keeps the delay discussion fair and less tense. In projects where the people are at each other's throats, some will use a delay issue to hurt the other side, and they will use contractual arguments as a tool.

• When a delay occurs, do the following:
  ○ Send a notice to the other party.
  ○ Start a chronological spreadsheet and type in all relevant activities as they happen. In the end, you will have a timeline of what happened.
  ○ Start a file for each issue. Save in that file all relevant documents: invoices, daily logs, invoices etc.
  ○ The above will be the backup for the claim.
  ○ Insert a delay activity in the schedule and tie it to the impacted activities.
  ○ Always show good faith by doing what you can to mitigate the delay impact. Fairness and honesty are the best policies.

• If you are an owner's representative responding to a delay claim, be fair. Look at the issues objectively and assess situations honestly. I have seen many owner's representatives who think that their automatic role is to find ways to deny the time extension request. By the same token, I have also seen contractors exploiting situations to claim more delay damages. Again, fairness and honesty are the best policies.

• If you are an owner, please conduct a thorough constructability review before you advertise the project for bidding. Remember the 5Ps: "Proper Planning Prevents Poor Performance."

Chapter 12
Construction Claims

The construction industry is very litigious due to the large amount of money that is involved, especially in tight bid environments. That is one of the main reasons why it is very important to understand all the above chapters and the need for proper documentation.

Claims are very costly, and you should try your best to avoid them. Here are some key steps to avoiding claims:

• Maintain fairness, honesty, and build good relationships on the project. I am sure you noticed that the recurring theme of the above chapters is the human nature element. Team members who get along fine will most likely resolve conflicts and avoid claims. I have seen so many claims escalate into costly litigations because of personality conflicts, egos, and bad attitudes. Some people will spend $300,000 on a $100,000 claim because they will not yield to the other side and will get them back if it was the last thing they do.

• Resolve conflict as soon as they arise. Don't let a dispute linger and drag unresolved.

• Maintain good documentation. The party with better documentation has an advantage in claims. So maintain good documentation, even if the project is going smoothly and the team has an excellent relationship. You never know what happens. But if it does, you are ready. If it doesn't, then no harm done.

• Understand your contractual obligation under the contract. You should always be on the right side of the contract. In other words, always do what you are supposed to do.

• Give timely notices in case of a problem, whether it is a delay event or a change order issue.

From reading the previous chapters, you already have an idea about the causes of claims. That includes change orders disputes, delay claims, defects, breaches of contract, and more.

What I would like to focus on is the process of claims and the various methods by which they are resolved.

Each contract has its specific process for claims. When you are managing a particular contract, you need to look up these clauses to know how to handle claims. Here is one of the procedures that I noticed to give you an example of what I'm talking about.

- If a dispute arises, the parties will try to resolve the issue at the project's team level.
- If the issue is not resolved, an upper-management meeting will be requested.
- If that doesn't work, the issue is now labeled a claim. In this case, the contractor has fifteen days to provide the claim with all the backup documents for consideration.
- The owner has ten days to respond after receipt of the contractor's claim.
- A senior-management meeting will be held to resolve the claim.
- If that doesn't work, both parties agree to resort to mediation.
- If the matter is not settled at mediation, the parties agree to a binding arbitration if the claim amount is less or equal to $500,000.
- If the claim is more than $500,000, then the parties will resort to court litigation.

That is one example, but will give you an idea. The key to any procedure is the timelines of when to file notices and papers; these are critical. Court litigation is very costly; that's why many parties resort to Alternate Dispute Resolution (ADR) means. Here are the most common methods.

Mediation

Mediation is an ADR process used to resolve conflicts between parties by trying to reach a mutually-agreeable resolution as an alternate to escalating the conflict to litigation in court, where the dynamics and relationship between the parties are completely different.

Preparing for mediation requires an understanding of the dynamics of the process and the mindset of the parties that elected, or were required by contract, to resort to mediation.

• In mediation, the parties of the dispute work together with the assistance of a mediator, an impartial person, to arrive at a mutually-agreeable solution.

• The mediation process protects the parties' interests and legal rights. In order to encourage the free exchange of information, all mediation sessions are confidential. That is not the case in standard litigation, where rules of evidence and discovery laws may apply differently.

• Resorting to mediation can be effective when the issues are complex or when there is a high level of emotion surrounding the dispute.

• Mediation is not a process to determine guilt or innocence. The mediator facilitates communications and exploration of various options that both parties can agree to. The mediator can't impose a solution.

• Legal representation is allowed but not mandatory.

• The parties participating in mediation are open to reaching a compromised resolution as opposed to litigated cases, where the parties talk to each other through their attorneys, and they feel they have reached a dead end with the other party. So mediation is obviously a less adversarial approach.

Understanding the above elements is important to developing the approach and strategy for the mediation as compared to other claim resolution methods. Please note a summary of my approach in preparing for and managing the mediation process:

Pre-mediation
• Each claim brings its own specific characteristics based on its elements. So I approach each claim on a case-by-case basis. My preliminary preparation for the mediation requires my understanding of the following:
  ○ Nature of the dispute.
  ○ Dollar value of the dispute.
  ○ The disputing parties.
  ○ The individuals involved in the dispute.
  ○ The history of the dispute.
  ○ The type of project.
  ○ All relevant facts that I can get. This includes interviewing the people involved to hear their sides of the story. It is helpful to

gain an understanding of both sides of the argument.

* I study the project's contract documents and conduct site visits as needed and any other necessary research to analyze the facts and arrive at a professional opinion regarding the conflict.

* Discuss my findings and conclusions with our client. Since this is mediation, we will explore with the client all possible solutions that may be presented by either side and discuss what the client is willing to accept or reject.

* Prepare an expert report, if required, in a particular case. In this case, we need to label the report "For Mediation Purposes Only." That will protect its confidentiality in case the mediation fails and the parties move on to litigation.

* We need to know the format of the mediation. Sometimes, the parties are seated in different rooms where the mediator goes back and forth, trying to bring both parties to a mutual agreement. In some cases, both parties and the mediator are in the same room. Sometimes it is a combination of the above.

* Based on the format, we will meet with the client and develop the negotiation strategy for the mediation.

* Assist in the mediator's selection process.

During Mediation

* The communication between the client and the owner's team during the mediation depend on the mediation format, as described above.

* A lead negotiator has to be assigned. This lead is the speaker of the group. He/she will be the main communicator with the mediator and/or the other team. This lead person will conduct the negotiation and privately consult with the client to discuss responses and decisions as needed.

* As a consultant to our client, one of my main responsibilities is to watch for our client's interest and provide professional advice and recommendations for the client's consideration based on the dialogue.

Arbitration

Arbitration is a form of dispute resolution. Arbitration is a private, judicial determination by an independent third party who is the arbitrator. The proceedings are similar to court proceedings, except they are conducted in designated centers equipped for these proceedings. But it is less costly than going through court.

Unlike mediation, the arbitrator's decision is final and binding. In mediation, the parties are there to negotiate and reach a common ground, resolution, or leave unresolved.

Arbitrators are commonly appointed by one of three means:

- Directly by the disputing parties.
- By existing tribunal members (for example, each side appoints one arbitrator, and then the arbitrators appoint a third).
- By an external party (for example, the court or an individual or institution nominated by the parties).

Arbitration is governed by state and federal law. Most states have provisions in their civil practice rules for arbitration. Many states have adopted the Uniform Arbitration Act, although some states have specific and individual rules for arbitration.

Advantages of Arbitration
- Choice of Arbitrator. Parties can choose a technical person as arbitrator if the dispute is of a technical nature, like construction, so that the evidence will be more readily understood.
- Efficiency. Arbitration can be completed faster than a standard court litigation. Final Decisions are final and binding. The locations can be selected to the convenience of the parties.
- Cost. One or both of the parties will pay for the arbitrator's services. Either way, it is still cheaper than going to court.
- No Appeal. Unless there is evidence of outright corruption or fraud, the award is binding and usually not appealable. Thus if the arbitrator makes a mistake or is simply an idiot, the losing party usually has no remedy.

Typical Steps in an Arbitration

- Initiating the Arbitration. The contract may dictate resorting to arbitration, or the parties may elect this process.
- Appointment of Arbitrator. Arbitrators may be appointed by the disputing parties or by an external party (for example, the court or an individual or institution nominated by the parties).
- Preliminary Meeting. It is a good idea to have a meeting between the arbitrator and the parties, along with their legal coun-

sels, to look over the dispute in question and discuss an appropriate process and timetable.

• Statement of Claim and Response. The claimant sets out a summary of the matters in dispute and the remedy sought in a statement of claim. This is needed to inform the respondent of what needs to be answered. It summarizes the alleged facts, but does not include the evidence through which facts are to be proved. The statement of response from the respondent is to admit or deny the claims. There may also be a counterclaim by the respondent, which in turn requires a reply from the claimant. These statements are called the pleadings. Their purpose is to identify the issues and avoid surprises.

• Discovery and Inspection. These are legal procedures through which the parties investigate background information. Each party is required to list all relevant documents, which are in their control. This is called discovery. Parties then "inspect" the discovered documents, and an agreed-upon selection of documents is prepared for the arbitrator.

• Interchange of Evidence. The written evidence is exchanged and given to the arbitrator for review prior to the hearing.

• Hearing. The hearing is a meeting in which the arbitrator listens to any oral statements, questioning of witnesses, and can ask for clarification of any information. Both parties are entitled to put forward their case and be present while the other side states theirs. A hearing may be avoided, however, if the issues can be dealt with entirely from the documents.

• Legal Submissions. The lawyers of both parties provide the arbitrator with a summary of their evidence and applicable laws. These submissions are made either orally at the hearing, or put in writing as soon as the hearing ends.

• Award. The arbitrator considers all the information and makes a decision. An award is written to summarize the proceedings and give the decisions. The award usually includes the arbitrator's reasons for the decision.

You see, arbitration is similar to a court proceeding, but less formal, faster, and less expensive.

The bottom line is *avoid claims at all costs*. Solve the problems as they arise at the project's level. Don't let an issue fester for a long time. The best cure is establishing fair and good relationships among the project's team members.

Conclusion

I hope I was able to elaborate effectively the various phases of projects, from inception to completion.

The book offers you a real-life insight on how to handle each of these phases. Keep this book handy as a reference to you as you move forward in your career of construction management.

Review the segments of the chapters labeled "Real-Life Gauge" and "Best Practices."

Appendix
Review of the Public Contract Code

Sections 4100-4114. This chapter may be cited as the "Subletting and Sub-contracting Fair Practices Act."

Section 4101. The legislature finds that the practices of bid shopping and bid peddling in connection with the construction, alteration, and repair of public improvements often result in poor quality of material and workmanship to the detriment of the public, deprive the public of the full benefits of fair competition among prime contractors and subcontractors, and lead to insolvencies, loss of wages to employees, and other evils.

Section 4103. Nothing in this chapter limits or diminishes any rights or remedies, either legal or equitable, which:

(a) An original or substituted subcontractor may have against the prime contractor, his or her successors or assigns.

(b) The state or any county, city, body politic, or public agency may have against the prime contractor, his or her successors or assigns, including the right to take over and complete the contract.

Section 4104. Any officer, department, board or commission taking bids for the construction of any public work or improvement shall provide in the specifications prepared for the work or improvement or in the general conditions under which bids will be received for the doing of the work incident to the public work or improvement that any person making a bid or offer to perform the work, shall, in his or her bid or offer, set forth:

(a) (1) The name and the location of the place of business of each subcontractor who will perform work or labor or render service to the prime contractor in or about the construction of the work or improvement, or a subcontractor licensed by the State of California who, under subcontract to the prime contractor, specially fabricates and installs a portion of the work or improvement according to detailed drawings contained in the plans and specifications, in an amount in excess of one-half of 1 percent of the prime contractor's total bid or, in the case of bids or offers for the construction of streets or highways, including bridges, in excess of one-half of 1 percent of the prime contractor's total bid or ten thousand dollars ($10,000), whichever is greater.

(2) (A) Subject to subparagraph (B), any information requested by the officer, department, board, or commission concerning any subcontractor who the prime contractor is required to list under this subdivision, other than the subcontractor's name and location of business, may be submitted by the prime contractor up to 24 hours after the deadline established by the officer, department, board, or commission for receipt of bids by prime contractors.

Section 20107. All bids for construction work shall be presented under sealed cover and shall be accompanied by one of the following forms of bidder's security:

(a) Cash.

(b) A cashier's check made payable to the school district.

(c) A certified check made payable to the school district.

(d) A bidder's bond executed by an admitted surety insurer, made payable to the school district. Upon an award to the lowest bidder, the security of an unsuccessful bidder shall be returned in a reasonable period of time, but in no event shall that security be held by the school district beyond 60 days from the time the award is made.

Section 10140. Public notice of a project shall be given by publication once a week for at least two consecutive weeks or once a week for more than two consecutive weeks if the longer period of advertising is deemed necessary by the department, as follows:

(a) In a newspaper of general circulation published in the county in which the project is located, or if located in more than one county, in such a newspaper in a county in which a major portion of the work is to be done.

(b) In a trade paper of general circulation published in San Francisco for projects located in County Group No. 1, as defined in Section 187 of the Streets and Highways Code, or in Los Angeles for projects located in County Group No. 2, as defined in said Section 187, devoted primarily to the dissemination of contract and building news among contracting and building materials supply firms. The department may publish the no-tice to bidders for a project in additional trade papers or newspapers of general circulation that it deems advisable.

Section 10141. The notice shall state the time and place for the receiving and opening of sealed bids, describing in general terms the work to be done and that the bids will be required for the entire project and for the performance of separate designated parts of the entire project, when the department determines that segregation is advisable.

Per section 1600-1601 any public entity may adopt methods and procedures to receive bids on public works or other contracts over the Internet, but only if no bid can be opened before the bid deadline and all bids can be verified as authentic.

Section 3300.

(a) Any public entity, as defined in Section 1100, the University of California, and the California State University shall specify the classification of the contractor's license, which a contractor shall possess at the time a contract is awarded. The specification shall be included in any plans prepared for a public project and in any notice inviting bids required pursuant to this code. This requirement shall apply only with respect to contractors who contract directly with the public entity.

(b) A contractor who is not awarded a public contract because of the failure of an entity, as defined in subdivision (a), to comply with that subdivision shall not receive damages for the loss of the contract.

Section 4104.5.

(a) The officer, department, board, or commission taking bids for construction of any public work or improvement shall specify in the bid invitation and public notice the place the bids of the prime contractors are to be received and the time by which they shall be received. The date and time shall be extended by no less than 72 hours if the officer, department, board, or commission issues any material changes, additions, or deletions to the invitation later than 72 hours prior to the bid closing. Any bids received after the time specified in the notice or any extension due to material changes shall be returned unopened.

(b) As used in this section, the term "material change" means a change with a substantial cost impact on the total bid as determined by the awarding agency.

(c) As used in this section, the term "bid invitation" shall include any documents issued to prime contractors that contain descriptions of the work to be bid or the content, form, or manner of submission of bids by bidders.

Section 4105. Circumvention by a general contractor who bids as a prime con-tractor of the requirement under Section 4104 for him or her to list his or her sub-contractors, by the device of listing another contractor who will in turn sublet portions constituting the majority of the work covered by the prime contract, shall be considered a violation of this chapter and shall subject that prime contractor to the penalties set forth in Sections 4110 and 4111.

Section 4106. If a prime contractor fails to specify a subcontractor or if a prime contractor specifies more than one subcontractor for the same portion of work to be performed under the contract in excess of one-half of 1 percent of the prime contractor's total bid, the prime contractor agrees that he or she is fully qualified to perform that portion himself or herself, and that the prime contractor shall perform that portion himself or herself. If after award of contract, the prime contractor subcontracts, except as provided for in Sections 4107 or 4109, any such portion of the work, the prime contractor shall be subject to the penalties named in Section 4111.

Section 4107. A prime contractor whose bid is accepted may not:

(a) Substitute a person as subcontractor in place of the subcontractor listed in the original bid, except that the awarding authority, or its duly authorized officer, may, except as otherwise provided in Section 4107.5, consent to the substitution of another person as a subcontractor in any of the following situations:

(1) When the subcontractor listed in the bid, after having had a reasonable opportunity to do so, fails or refuses to execute a written contract for the scope of work specified in the subcontractor's bid and at the price specified in the subcontractor's bid, when that written contract, based upon the general terms, conditions, plans, and specifications for the project involved or the terms of that subcontractor's written bid, is presented to the subcontractor by the prime contractor.

(2) When the listed subcontractor becomes bankrupt or insolvent.

(3) When the listed subcontractor fails or refuses to perform his or her subcontract.

(4) When the listed subcontractor fails or refuses to meet the bond requirements of the prime contractor as set forth in Section 4108.

(5) When the prime contractor demonstrates to the awarding authority, or its duly authorized officer, subject to the further provi-

sions set forth in Section 4107.5, that the name of the subcontractor was listed as the result of an inadvertent clerical error.

(6) When the listed subcontractor is not licensed pursuant to the Contractor's License Law.

(7) When the awarding authority, or its duly authorized officer, determines that the work performed by the listed subcontractor is substantially unsatisfactory and not in substantial accordance with the plans and specifications, or that the subcontractor is substantially delaying or disrupting the progress of the work.

(8) When the listed subcontractor is ineligible to work on a public works project pursuant to Section 1777.1 or 1777.7 of the Labor Code.

(9) When the awarding authority determines that a listed subcontractor is not a responsible contractor. Prior to approval of the prime contractor's request for the substitution, the awarding authority, or its duly authorized officer, shall give notice in writing to the listed subcontractor of the prime contractor's request to substitute and of the reasons for the request. The notice shall be served by certified or registered mail to the last known address of the subcontractor. The listed subcontractor who has been so notified has five working days within which to submit written objections to the substitution to the awarding authority. Failure to file these written objections constitutes the listed subcontractor's consent to the substitution. If written objections are filed, the awarding authority shall give notice in writing of at least five working days to the listed subcontractor of a hearing by the awarding authority on the prime contractor's request for substitution.

(b) Permit a subcontract to be voluntarily assigned or transferred or allow it to be performed by anyone other than the original subcontractor listed in the original bid, without the consent of the awarding authority, or its duly authorized officer.

(c) Other than in the performance of "change orders" causing changes or deviations from the original contract, sublet or subcontract any portion of the work in excess of one-half of 1 percent of the prime contractor's total bid as to which his or her original bid did not designate a subcontractor.

Section 4107.2. No subcontractor listed by a prime contractor under Section 4104 as furnishing and installing carpeting, shall voluntarily sublet his or her subcontract with respect to any portion of the labor to be performed unless he or she specified the subcontractor in his or her bid for that subcontract to the prime contractor.

Section 4107.5. The prime contractor as a condition to assert a claim of inadvertent clerical error in the listing of a subcontractor shall within two working days after the time of the prime bid opening by the awarding authority give written no-tice to the awarding authority and copies of that notice to both the subcontractor he or she claims to have listed in error and the intended subcontractor who had bid to the prime contractor prior to bid opening.

Any listed subcontractor who has been notified by the prime contractor in accordance with this section as to an inadvertent clerical error shall be allowed six working days from the time of the prime bid opening within which to submit to the awarding authority and to the prime contractor written objection to the prime contractor's claim of inadvertent clerical error. Failure of the listed subcontractor to file the written notice within the six working days shall be primary evidence of his or her agreement that an inadvertent clerical error was made. The awarding authority shall, after a public hearing as provided in Section 4107 and in the absence of compelling reasons to the contrary, consent to the substitution of the intended subcontractor:

(a) If (1) the prime contractor, (2) the subcontractor listed in error, and (3) the intended subcontractor each submit an affidavit to the awarding authority along with such additional evidence as the parties may wish to submit that an inadvertent clerical error was in fact made, provided that the affidavits from each of the three parties are filed within eight working days from the time of the prime bid opening, or

(b) If the affidavits are filed by both the prime contractor and the intended subcontractor within the specified time but the subcontractor whom the prime contractor claims to have listed in error does not submit within six working days, to the awarding authority and to the prime contractor, writ-ten objection to the prime contractor's claim of inadvertent clerical error as provided in this section.

If the affidavits are filed by both the prime contractor and the intended subcontractor but the listed subcontractor has, within six working days from the time of the prime bid opening, submitted to the awarding authority and to the prime contractor written objection to the prime contractor's claim of inadvertent clerical error, the awarding authority shall investigate the claims of the parties and shall hold a public hearing as provided in Section 4107 to determine the validity of those claims. Any determination made shall be based on the facts contained in the declarations submitted under penalty of perjury by all three parties and supported by testimony under oath and subject to cross-examination. The awarding authority may, on its own motion or that of any other party, admit testimony of other con-tractors, any bid registries or depositories, or any other party in possession of facts which may have a bearing on the decision of the awarding authority.

Section 4107.7. If a contractor who enters into a contract with a public entity for investigation, removal or remedial action, or disposal relative to the release or presence of a hazardous material or hazardous waste fails to pay a subcontractor registered as a hazardous waste hauler pursuant to Section 25163 of the Health and Safety Code within 10 days after the investigation, removal or remedial action, or disposal is completed, the subcontractor may serve a stop notice upon the public entity in accordance with Chapter 4 (commencing with Section 3179) of Title 15 of Part 4 of Division 3 of the Civil Code.

Section 4108.

(a) It shall be the responsibility of each subcontractor submitting bids to a prime contractor to be prepared to submit a faithful performance and payment bond or bonds if so requested by the prime contractor.

(b) In the event any subcontractor submitting a bid to a prime contractor does not, upon the request of the prime contractor and at the expense of the prime contractor at the established charge or premium therefore, furnish to the prime contractor a bond or bonds issued by an admitted surety wherein the prime contractor shall be named the oblige, guaranteeing prompt and faithful performance of the subcontract and the payment of all claims for labor and materials furnished or used in and about the work to be done and performed under the subcontract, the prime contractor may reject the bid and make a substitution of another subcontractor subject to Section 4107.

(c) (1) The bond or bonds may be required under this section only if the prime contractor in his or her written or published request for subbids clearly specifies the amount and requirements of the bond or bonds.

(d) If the expense of the bond or bonds required under this section is to be borne by the subcontractor, that requirement shall also be specified in the prime contractor's written or published request for subbids.

(e) The prime contractor's failure to specify bond requirements, in accordance with this subdivision, in the written or published request for subbids shall preclude the prime contractor from imposing bond requirements under this section.

Section 4109. Subletting or subcontracting of any portion of the work in excess of one-half of 1 percent of the prime contractor's total bid as to which no subcontractor was designated in the original bid shall only be permitted in cases of public emergency or necessity, and then only after a finding reduced to writing as a public record of the awarding authority setting forth the facts constituting the emergency or necessity.

Section 4110. A prime contractor violating any of the provisions of this chapter violates his or her contract and the awarding authority may exercise the option, in its own discretion, of (1) canceling his or her contract or (2) assessing the prime contractor a penalty in an amount of not more than 10 percent of the amount of the subcontract involved, and this penalty shall be deposited in the fund out of which the prime contract is awarded. In any proceedings under this section the prime contractor shall be entitled to a public hearing and to five days' notice of the time and place thereof.

Section 4111. Violation of this chapter by a licensee under Chapter 9 (commencing with Section 7000) of Division 3 of the Business and Professions Code constitutes grounds for disciplinary action by the Contractors State License Board, in addition to the penalties prescribed in Section 4110.

Section 4112. The failure on the part of a contractor to comply with any provision of this chapter does not constitute a defense to the contractor in any action brought against the contractor by a subcontractor.

Section 4113. As used in this chapter, the word "subcontractor" shall mean a contractor, within the meaning of the provisions of Chapter 9 (commencing with Section 7000) of Division 3 of the Business and Professions Code, who contracts directly with the prime contractor.

"Prime contractor" shall mean the contractor who contracts directly with the awarding authority.

Section 4114. The county board of supervisors, when it is the awarding authority, may delegate its functions under Sections 4107 and 4110 to any officer designated by the board. The authorized officer shall make a written recommendation to the board of supervisors. The board of supervisors may adopt the recommendation without further notice or hearing, or may set the matter for a de novo hearing before the board.

Section 5100 - 5110

(a) A bidder shall not be relieved of the bid unless by consent of the awarding authority nor shall any change be made in the bid because of mistake, but the bidder may bring an action against the public entity in a court of competent jurisdiction in the county in which the bids were opened for the recovery of the amount forfeited, without interest or costs. If the plaintiff fails to recover judgment, the plaintiff shall pay all costs incurred by the public entity in the suit, including a reasonable attorney's fee to be fixed by the court.

(b) If an awarding authority for the state consents to relieve a bidder of a bid because of mistake, the authority shall prepare a report in writing to document the facts establishing the existence of each element required by Section 5103. The report shall be available for inspection as a public record. In the case of the University of California or a California State University, the report shall be filed with the regents and the trustees, respectively, and shall be available as a public record.

Section 5102. The complaint shall be filed, and summons served on the director of the department or the chief of the division or other head of the public entity under which the work is to be performed or an appearance made, within 90 days after the opening of the bid; otherwise, the action shall be dismissed.

Section 5103. The bidder shall establish to the satisfaction of the court that:

(a)   A mistake was made.

(b)   He or she gave the public entity written notice within five working days, excluding Saturdays, Sundays, and state holidays, after the opening of the bids of the mistake, specifying in the notice in detail how the mistake occurred.

(c)   The mistake made the bid materially different than he or she intended it to be.

(d)   The mistake was made in filling out the bid and not due to error in judgment or to carelessness in inspecting the site of the work, or in reading the plans or specifications.

Section 5104. Other than the notice to the public entity, no claim is required to be filed before bringing the action.

Section 5105. A bidder who claims a mistake or who forfeits his or her bid security shall be prohibited from participating in further bidding on the project on which the mistake was claimed or security forfeited.

Section 5106. If the public entity deems it is for its best interest, it may, on refusal or failure of the successful bidder to execute the contract, award it to the second lowest bidder. If the second lowest bidder fails or refuses to execute the contract, the public entity may likewise award it to the third lowest bidder.

On the failure or refusal of the second or third lowest bidder - to whom a contract is so awarded to execute it, his or her bidder's security shall be likewise forfeited.

Per Section 7103. (a) Every original contractor to who is awarded a contract by a state entity, as defined in subdivision (d), involving an expenditure in excess of five thousand dollars ($5,000) for any public work shall, before entering up the performance of the work, file a payment bond with and approved by the officer or state entity by who the contract was awarded. The bond shall be in a sum not less than one hundred percent of the total amount payable by the terms of the contract. The state entity shall state in its call for bids for any contract that a payment bond is required in the case of such an expenditure.